SpringerBriefs in Physics

SpringerBriefs in Physics are a series of slim high-quality publications encompassing the entire spectrum of physics. Manuscripts for SpringerBriefs in Physics will be evaluated by Springer and by members of the Editorial Board. Proposals and other communication should be sent to your Publishing Editors at Springer.

Featuring compact volumes of 50 to 125 pages (approximately 20,000–45,000 words), Briefs are shorter than a conventional book but longer than a journal article. Thus, Briefs serve as timely, concise tools for students, researchers, and professionals.

Typical texts for publication might include:

- A snapshot review of the current state of a hot or emerging field
- A concise introduction to core concepts that students must understand in order to make independent contributions
- An extended research report giving more details and discussion than is possible in a conventional journal article
- A manual describing underlying principles and best practices for an experimental technique
- An essay exploring new ideas within physics, related philosophical issues, or broader topics such as science and society

Briefs allow authors to present their ideas and readers to absorb them with minimal time investment.

Briefs will be published as part of Springer's eBook collection, with millions of users worldwide. In addition, they will be available, just like other books, for individual print and electronic purchase.

Briefs are characterized by fast, global electronic dissemination, straightforward publishing agreements, easy-to-use manuscript preparation and formatting guidelines, and expedited production schedules. We aim for publication 8–12 weeks after acceptance.

More information about this series at http://www.springer.com/series/8902

Leon Karpa

Trapping Single Ions and Coulomb Crystals with Light Fields

 Springer

Leon Karpa
Institute of Physics
University of Freiburg
Freiburg im Breisgau
Baden-Württemberg, Germany

ISSN 2191-5423   ISSN 2191-5431   (electronic)
SpringerBriefs in Physics
ISBN 978-3-030-27715-4   ISBN 978-3-030-27716-1   (eBook)
https://doi.org/10.1007/978-3-030-27716-1

This Springer imprint is published by the registered company Springer Nature Switzerland AG.
The registered company address is: Gewerbestrasse 11, 6330 Cham, Switzerland

This work is dedicated to my family.

Preface

The scope of this book is to provide insights into the recently emerged field of optical trapping of ions. Ever since the groundbreaking introduction of light fields as tools for exerting trapping forces on matter in 1970 by Ashkin [1], optical dipole traps have been used with remarkable success in several fields of research, most notably in atomic physics, where they have enabled an unprecedented level of control over neutral atoms and molecules at the level of both quantum ensembles [2] and individual particles [3, 4].

Another tremendously influential development in physics was enabled by the invention of radiofrequency traps pioneered by Paul [5] and Dehmelt, allowing for the confinement of single atomic ions demonstrated by Wineland [6] decades before this was realized on the level of individual neutral atoms in optical traps. Ions provide a number of properties that render them ideal for state-of-the-art applications in several cutting-edge areas of contemporary research ranging from metrology [7], quantum simulation and computation [8] to ultracold chemistry [9].

However, it was found recently that in some situations, it is highly advantageous to confine ions without employing any radiofrequency-based Paul traps or strong external magnetic fields in Penning traps, e.g., when investigating the interaction of neutral atoms and ions in the regime of ultralow interaction energies [10]. Adapting optical traps for ions [11] is a promising way to approach such scenarios, and the focus of this work is to present a comprehensive overview of the background and concepts behind this technique as well as to discuss the currently achievable level of control, encountered limitations, and perspectives for future applications.

Chapter 1 is intended to provide a context of the described experiments within the field of atomic physics. It highlights the advantages of the currently used techniques applied for trapping and manipulating ions in comparison with the features granted by optical traps for neutral atoms.

Chapter 2 lays out the general concepts, requirements, and methodology for realizations of optical trapping of ions.

Chapter 3 describes how the tools and methods presented in the previous chapter can be used to confine single atomic ions in optical traps. Individual developments achieved in the last years are highlighted by presenting and discussing milestone experiments [12–15]. The aim of this chapter is to provide an understanding of the current performance granted by this approach, and its prospective possibilities for further-going applications in the near future. It concludes with a discussion of the identified limitations and strategies for future improvement.

Chapter 4 is focused on experiments reaching beyond the established optical traps for single ions. It provides an in-depth view on the recently demonstrated extension of the presented methods to the case of ion Coulomb crystals [16].

Chapter 5 highlights the key features of the concepts discussed in this book, summarizes the main features of optical traps for ions, and provides an outlook for future applications building on the capabilities afforded by these novel tools.

Freiburg im Breisgau, Germany   Leon Karpa

References

1. A. Ashkin, Acceleration and trapping of particles by radiation pressure. Phys. Rev. Lett. **24**, 156–159 (1970)
2. I. Bloch, J. Dalibard, W. Zwerger, Many-body physics with ultracold gases. Rev. Mod. Phys. **80**, 885–964 (2008)
3. M. Endres et al., Atom-by-atom assembly of defect-free one-dimensional cold atom arrays. Science **354**, 1024–1027 (2016)
4. D. Barredo et al., Synthetic three-dimensional atomic structures assembled atom by atom. Nature **561**, 79–82 (2018)
5. W. Paul, Electromagnetic traps for charged and neutral particles. Rev. Mod. Phys. **62**, 531–540 (1990)
6. D.J. Wineland, Nobel lecture: superposition, entanglement, and raising Schrödinger's cat. Rev. Mod. Phys. **85**, 1103–1114 (2013)
7. T. Rosenband et al., Frequency ratio of Al and Hg single-ion optical clocks metrology at the 17th decimal place. Science **319**, 1808–1812 (2008)
8. R. Blatt, D. Wineland, Entangled states of trapped atomic ions. Nature **453**, 1008–1015 (2008)
9. R. Côté, V. Kharchenko, M.D. Lukin, Mesoscopic molecular ions in Bose-Einstein condensates. Phys. Rev. Lett. **89**, 093001 (2002)
10. A. Härter, J.H. Denschlag, Cold atom–ion experiments in hybrid traps. Contemp. Phys. **55**, 33–45 (2014)
11. T. Schaetz, Trapping ions and atoms optically. J. Phys. B At. Mol. Opt. Phys. **50**, 102001 (2017)
12. C. Schneider et al., Optical trapping of an ion. Nat. Photonics **4**, 772–775 (2010)
13. M. Enderlein et al., Single ions trapped in a one-dimensional optical lattice. Phys. Rev. Lett. **109**, 233004 (2012)
14. T. Huber et al., A far-off-resonance optical trap for a Ba^+ ion. Nat. Commun. **5**, 5587 (2014)
15. A. Lambrecht et al., Long lifetimes and effective isolation of ions in optical and electrostatic traps. Nat. Photonics **11**, 704–707 (2017)
16. J. Schmidt et al., Optical trapping of ion Coulomb crystals. Phys. Rev. X **8**, 021028 (2018)

Acknowledgments

The projects described in this work have received funding from the European Research Council (ERC) under the European Union's Horizon 2020 research and innovation program (Grant No. 648330). The author is grateful for the financial support from the Alexander von Humboldt Foundation and Marie Skłodowska-Curie Actions.

I am indebted to all my colleagues who have contributed to several generations of optical trapping experiments. In particular, it was my privilege to work side by side with T. Huber, J. Schmidt, A. Lambrecht, and of course T. Schätz who has conceived and initiated the experiments on optical ion trapping. I would especially like to thank Vladan Vuletić for introducing me to the field, our numerous fruitful discussions, as well as his prudent advice and support.

Contents

About the Author

Leon Karpa studied physics at the Eberhard Karls Universität Tübingen in Germany where he earned a Diploma in Physics. He continued his research in the field of experimental quantum optics at Rheinische Friedrich-Wilhems-Universität Bonn in Germany where he was awarded a doctorate in 2010. He then accepted a Feodor Lynen Research Fellowship of the Alexander von Humboldt Foundation at the Massachusetts Institute of Technology, USA, where he pursued his research interests in two then arising promising topics related to the fields of cavity quantum electrodynamics and quantum information processing: collective ion-photon interfaces and optical trapping of ions in hybrid traps. In 2013, he was awarded with a Return Fellowship of the Alexander von Humboldt Foundation at the Albrecht-Ludwigs-Univesität Freiburg to continue his work on optical trapping of ions and atoms, as well as their interaction in the ultracold regime. His recent awards include a Marie Skłodowska-Curie Fellowship of the European Union at the Freiburg Institute of Advanced Studies.

Chapter 1
Introduction

Over the past decades, optical forces have proved to be a pivotal instrument for confinement and manipulation of atoms [1–3]. Their use has enabled highly versatile trap geometries while at the same time preserving the capability to investigate interactions of atomic gases and molecules at ultralow temperatures [4]. Optical traps are now routinely used to trap and control individual neutral atoms and atomic ensembles on the nanoscale with sub-wavelength resolution, e.g. in the Mott insulator regime [5] or in the demonstrated realizations of quantum gas microscopes [6, 7] (Fig. 1.1(a)). Recently these techniques have been once more tremendously improved by refining the control over combinations of several optical fields and making use of the available degrees of freedom such as geometry, intensity, polarizations, and relative phases [4, 8, 9]. Consequently, this progress has led to the realization of superlattices [10], and Kagomé lattices [11], as well as to the assembly of linear [12], two-dimensional [13] and even three-dimensional arrays [14] of atoms (Fig. 1.1(b)).

For several decades, the exquisite degree of control over individual atoms has been one of most prominent characteristic features in the field of ion trapping, a vastly successful and rapidly expanding area of atomic physics [16, 17]. In this field of research, individual atoms are typically confined in radiofrequency or Penning traps, affording preparation, operation, and detection fidelities approaching unity and storage times on the order of months or years. Motional and internal electronic degrees of freedom of atomic ions can be manipulated on the quantum level with individual addressability and exceptionally high operation fidelities in excess of 99% [18–20], making trapped ions one of the most promising and performant platforms for quantum information processing (QIP), quantum simulations, and metrology to date. Long-range interactions that are difficult to establish in experiments with neutral atoms allow for using phonon modes of Coulomb crystals as a bus for generating entanglement between ions in linear chains [21].

While Paul traps have proved to be very successful or leading platforms for applications in many fields of contemporary research, some outstanding challenges

L. Karpa, *Trapping Single Ions and Coulomb Crystals with Light Fields*, SpringerBriefs in Physics, https://doi.org/10.1007/978-3-030-27716-1_1

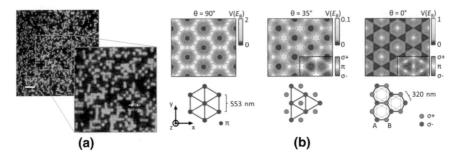

Fig. 1.1 Optical trapping potentials in experiments with neutral atoms: (**a**) individual atoms imaged within the optical potential of a standing wave (courtesy of W. Bakr, taken from [6]) and (**b**) versatile optical trapping geometries (from left to right: triangular, polarization, and honeycomb lattice) obtained by using a superposition of three laser beams and controlling their respective polarizations (courtesy of K. Sengstock, taken from [15]). Reprinted figure with permission from [15]. Copyright 2019 by the American Physical Society

Fig. 1.2 (**a**) Linear Paul trap with segmented electrodes. The diagonal spacing between the electrodes is 18 mm. (**b**) False-color fluorescence image of a three-dimensional ion Coulomb crystal. The spacing between the ions is determined by the choice of the trap parameters. Typically, the ions are separated by approximately 20 μm allowing for individual addressability

closely related to limitations in scalability have emerged (Fig. 1.2). The generation of nanoscale potentials with Paul traps remains another challenge despite of rapid advances achieved with planar trapping geometries. An intrinsic property of such radiofrequency (rf) traps is the driven motion at the modulation frequency which is superimposed with the secular motion of ions as illustrated in Fig. 1.3. While this so-called micromotion can be neglected in most applications based on the manipulation of common mode excitations within linear ion crystals, it can impose a fundamental limitation on other experiments sensitive to kinetic energy [22, 23]. A prominent example is the rapidly growing field of ion–atom interactions [24–31] where the presence of driven electromagnetic fields typically limits the accessible collision energies to the range of 1 mK [23, 32]. While some approaches exploiting the favorable kinematic properties of specific combinations of neutral atom and atomic ion species confined in conventional Paul traps exist [22, 29] and seem promising for

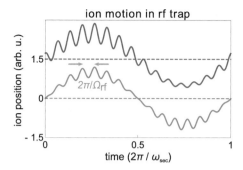

Fig. 1.3 Motion of an ion in a radiofrequency trap. The oscillation at a frequency ω_{sec} in the quasi-harmonic trapping potential is superimposed with a rapid oscillation at the modulation frequency of the trap, Ω_{rf}. In the presence of any stray electric field the amplitude of this micromotion is non-zero at all times (upper trace)

accessing a range of collision energies that could approach the quantum interaction regime, the generalization to either generic atom-ion combinations or to experiments involving more than a single ion or a linear chain remains an outstanding challenge in the field and stands to benefit from alternative ways to avoid micromotion. One recently demonstrated approach along these lines relies on the preparation of highly excited Rydberg states within a Bose–Einstein condensate, which then behave as ions immersed in an atomic bath [28]. So far the achieved kinetic energies of the Rydberg atoms correspond to temperatures on the order of microkelvin, and the sensitivity of the system to stray electric fields as well as the comparably short lifetime of the Rydberg excitations remain challenging.

In view of these considerations, a naturally seeming approach for addressing the issues highlighted above would be to use the powerful set of tools and methods provided by optical potentials and combine it with the advantage granted by ions and Coulomb crystals as a platform for experimental investigations. The recently emerged field of optical traps for ions aims at harvesting exactly the potential of this synergetic combination [33]. Dynamically applied optical forces are instrumental in quantum information processing with ions [17, 34] as a route towards mediating state-dependent interactions [21, 35–37]. Static optical potentials have been used to study anomalous ion diffusion [38]. Despite the broad use of optical forces and potentials for the manipulation of ions, it was only in 2010 that they have been successfully used to demonstrate optical trapping of a single ion [39] in absence of any rf fields. Since then the employed techniques and methods have undergone a continuous development in terms of performance and feasibility culminating in a series of experiments [33, 40–45]. These investigations will be in the focus of the following discussion aiming to provide a dissemination of the basic concepts behind this approach as well as its applicability to the outstanding experimental issues outlined in the previous paragraphs.

Another closely related class of experiments is based on a combination of radiofrequency traps with optical standing waves [46–48]. This approach provides composite trapping potentials with wavelength-scale periodicity whose implementation in conventional, purely rf-driven, traps including planar chip technology, to this day remains a highly demanding task exceeding the state-of-the-art. It is worth pointing out that combining several established trapping techniques in order to obtain a system with a unique set of advantageous properties has been a very successful strategy in the past. For example, a combination of linear Paul traps and Penning traps allows for a suppression of the magnetron frequency and can lead to an improved performance of mass spectrometers [49]. In the more recent case, the obtained nanoscale corrugation of the effective potential in one dimension is the unique feature of the hybrid setups mentioned above. They turned out to be a nearly ideal platform for investigating theoretically treated phenomena such as structural transitions of one-dimensional ion chains [50, 51] and have recently lead to the first realization of friction models on the level of a few atoms [52, 53]. Although in this hybrid trapping approach the presence of micromotion still hinders the straightforward extension to higher dimensional systems, in the future, radiofrequency-free trapping potentials realized in optical traps could open up a novel field of studies shedding light on structural phase transitions between one-, two-, and three-dimensional ordering close to a quantum critical point [54–57]. This however requires the capability to prepare such higher dimensional Coulomb crystals in the quantum regime which is another outstanding challenge in contemporary atomic physics.

On a similar note, the availability of quasi-static and state-dependent potentials confining 2d ion crystals could also provide a platform for novel quantum simulations [41, 58], e.g. the study of fracton models [59] or contribute to the field of quantum information processing by utilizing the state-dependent nature of optical potentials to realize fast gate operations as an alternative to approaches based on Rydberg ions [60–62] or fast Raman transitions [63]. Lastly, ions in optical potentials may be a used as a nanoscale probe with implications in the field of metrology and sensing of ambient fields.

The aim of the following sections is to present an overview of the recent advances in the field of optical trapping of ions and Coulomb crystals as well as to provide a detailed description of the prerequisites, experimental techniques, limitations, and prospects of this approach for such envisioned applications.

References

1. W.D. Phillips, Nobel lecture: laser cooling and trapping of neutral atoms. Rev. Mod. Phys. **70**, 721–741 (1998). https://doi.org/10.1103/RevModPhys.70.721
2. S. Chu, Nobel lecture: the manipulation of neutral particles. Rev. Mod. Phys. **70**, 685–706 (1998). https://doi.org/10.1103/RevModPhys.70.685
3. C.N. Cohen-Tannoudji, Nobel lecture: manipulating atoms with photons. Rev. Mod. Phys. **70**, 707–719 (1998). https://doi.org/10.1103/RevModPhys.70.707
4. I. Bloch, J. Dalibard, W. Zwerger, Many-body physics with ultracold gases. Rev. Mod. Phys. **80**, 885–964 (2008). https://doi.org/10.1103/RevModPhys.80.885

5. I. Bloch, Ultracold quantum gases in optical lattices. Nat. Phys. **1**(1), 23–30 (2005). https://doi.org/10.1038/nphys138

6. W.S. Bakr, J.I. Gillen, A. Peng, S. Fölling, M. Greiner, A quantum gas microscope for detecting single atoms in a Hubbard-regime optical lattice. Nature **462**(7269), 74–77 (2009). ISSN 0028-0836. https://doi.org/10.1038/nature08482. http://www.nature.com/articles/nature08482

7. J.F. Sherson, C. Weitenberg, M. Endres, M. Cheneau, I. Bloch, S. Kuhr, Single-atom-resolved fluorescence imaging of an atomic Mott insulator. Nature **467**(7311), 68–72 (2010). ISSN 0028-0836. https://doi.org/10.1038/nature09378. http://www.nature.com/articles/nature09378

8. D. Greif, G. Jotzu, M. Messer, R. Desbuquois, T. Esslinger, Formation and dynamics of antiferromagnetic correlations in tunable optical lattices. Phys. Rev. Lett. **115**, 260401, (2015). https://doi.org/10.1103/PhysRevLett.115.260401

9. O. Morsch, M. Oberthaler, Dynamics of Bose-Einstein condensates in optical lattices. Rev. Mod. Phys. **78**(1), 179–215 (2006). ISSN 00346861. https://doi.org/10.1103/RevModPhys.78.179

10. J. Kangara, C. Cheng, S. Pegahan, I. Arakelyan, J.E. Thomas, Atom pairing in optical superlattices. Phys. Rev. Lett. **120**(8), 083203 (2018). ISSN 0031-9007. https://doi.org/10.1103/PhysRevLett.120.083203. http://arxiv.org/abs/1709.08484

11. G.-B. Jo, J. Guzman, C.K. Thomas, P. Hosur, A. Vishwanath, D.M. Stamper-Kurn, Ultracold atoms in a tunable optical kagome lattice. Phys. Rev. Lett. **108**, 045305 (2012). https://doi.org/10.1103/PhysRevLett.108.045305

12. M. Endres, H. Bernien, A. Keesling, H. Levine, E.R. Anschuetz, A. Krajenbrink, C. Senko, V. Vuletic, M. Greiner, M.D. Lukin, Atom-by-atom assembly of defect-free one-dimensional cold atom arrays. Science **354**(6315), 1024–1027 (2016). ISSN 0036-8075. https://doi.org/10.1126/science.aah3752.

13. D. Barredo, S. de Léséleuc, V. Lienhard, T. Lahaye, A. Browaeys, An atom-by-atom assembler of defect-free arbitrary two-dimensional atomic arrays. Science **354**(6315), 1021–1023 (2016)

14. D. Barredo, V. Lienhard, S. de Léséleuc, T. Lahaye, A. Browaeys, Synthetic three-dimensional atomic structures assembled atom by atom. Nature **561**(7721), 79–82 (2018). ISSN 0028-0836. http://dx.doi.org/10.1038/s41586-018-0450-2

15. D.-S. Lühmann, O. Jürgensen, M. Weinberg, J. Simonet, P. Soltan-Panahi, K. Sengstock, Quantum phases in tunable state-dependent hexagonal optical lattices. Phys. Rev. A **90**, 013614 (2014). https://doi.org/10.1103/PhysRevA.90.013614

16. W. Paul, Electromagnetic traps for charged and neutral particles. Rev. Mod. Phys. **62**(3), 531–540 (1990). https://doi.org/10.1103/revmodphys.62.531

17. D.J. Wineland, Nobel lecture: superposition, entanglement, and raising Schrödinger's cat. Rev. Mod. Phys. **85**(3), 1103–1114 (2013). https://doi.org/10.1103/revmodphys.85.1103

18. C. Monroe, J. Kim, Scaling the ion trap quantum processor. Science **339**(6124), 1164–1169 (2013). ISSN 0036-8075. https://doi.org/10.1126/science.1231298

19. J.P. Gaebler, T.R. Tan, Y. Lin, Y. Wan, R. Bowler, A.C. Keith, S. Glancy, K. Coakley, E. Knill, D. Leibfried, D.J. Wineland, High-fidelity universal gate set for $^9Be^+$ ion qubits. Phys. Rev. Lett. **117**(6), 060505 (2016). ISSN 0031-9007. https://doi.org/10.1103/PhysRevLett.117.060505

20. C.J. Ballance, T.P. Harty, N.M. Linke, M.A. Sepiol, D.M. Lucas, High-fidelity quantum logic gates using trapped-ion hyperfine qubits. Phys. Rev. Lett. **117**(6), 060504 (2016). ISSN 0031-9007. https://doi.org/10.1103/PhysRevLett.117.060504

21. R. Blatt, C.F. Roos, Quantum simulations with trapped ions. Nat. Phys. **8**(4), 277–284 (2012). https://doi.org/10.1038/nphys2252

22. M. Cetina, A.T. Grier, V. Vuletić, Micromotion-induced limit to atom-ion sympathetic cooling in Paul traps. Phys. Rev. Lett. **109**, 253201 (2012). https://doi.org/10.1103/PhysRevLett.109.253201

23. M. Tomza, K. Jachymski, R. Gerritsma, A. Negretti, T. Calarco, Z. Idziaszek, P.S. Julienne, Cold hybrid ion-atom systems. arXiv (2017), pp. 1–59. http://arxiv.org/abs/1708.07832

24. A.T. Grier, M. Cetina, F. Oručević, V. Vuletić, Observation of cold collisions between trapped ions and trapped atoms. Phys. Rev. Lett. **102**, 223201 (2009). https://doi.org/10.1103/PhysRevLett.102.223201

25. S. Schmid, A. Härter, J.H. Denschlag, Dynamics of a cold trapped ion in a Bose-Einstein condensate. Phys. Rev. Lett. **105**, 133202 (2010). https://doi.org/10.1103/PhysRevLett.105. 133202

26. C. Zipkes, L. Ratschbacher, S. Palzer, C. Sias, M. Köhl, Hybrid quantum systems of atoms and ions. J. Phys. Conf. Ser. **264**(1), 012019 (2011). http://stacks.iop.org/1742-6596/264/i=1/a=012019

27. Z. Meir, T. Sikorsky, R. Ben-shlomi, N. Akerman, Y. Dallal, R. Ozeri, Dynamics of a ground-state cooled ion colliding with ultracold atoms. Phys. Rev. Lett. **117**, 243401 (2016). https://doi.org/10.1103/PhysRevLett.117.243401

28. K.S. Kleinbach, F. Engel, T. Dieterle, R. Löw, T. Pfau, F. Meinert, Ionic impurity in a Bose-Einstein condensate at submicrokelvin temperatures. Phys. Rev. Lett. **120**, 193401 (2018). https://doi.org/10.1103/PhysRevLett.120.193401

29. H.A. Fürst, N.V. Ewald, T. Secker, J. Joger, T. Feldker, R. Gerritsma, Prospects of reaching the quantum regime in Li–Yb+ mixtures. J. Phys. B At. Mol. Opt. Phys. **51**(19), 195001 (2018). ISSN 0953-4075. https://doi.org/10.1088/1361-6455/aadd7d. http://stacks.iop.org/0953-4075/51/i=19/a=195001?key=crossref.ec02c99f20b258fff1b50c2d303e694f

30. S. Dutta, R. Sawant, S.A. Rangwala, Collisional cooling of light ions by cotrapped heavy atoms. Phys. Rev. Lett. **118**, 113401 (2017). https://doi.org/10.1103/PhysRevLett.118.113401

31. S. Haze, M. Sasakawa, R. Saito, R. Nakai, T. Mukaiyama, Cooling dynamics of a single trapped ion via elastic collisions with small-mass atoms. Phys. Rev. Lett. **120**(4), 043401 (2018). ISSN 0031-9007. https://doi.org/10.1103/PhysRevLett.120.043401

32. A. Härter, J.H. Denschlag, Cold atom–ion experiments in hybrid traps. Contemp. Phys. **55**(1), 33–45 (2014). https://doi.org/10.1080/00107514.2013.854618

33. T. Schaetz, Trapping ions and atoms optically. J. Phys. B At. Mol. Opt. Phys. **50**(10), 102001 (2017)

34. D. Leibfried, R. Blatt, C. Monroe, D. Wineland, Quantum dynamics of single trapped ions. Rev. Mod. Phys. **75**, 281–324 (2003). https://doi.org/10.1103/RevModPhys.75.281

35. C. Monroe, D.M. Meekhof, B.E. King, S.R. Jefferts, W.M. Itano, D.J. Wineland, P. Gould, Resolved-sideband Raman cooling of a bound atom to the 3D zero-point energy. Phys. Rev. Lett. **75**(22), 4011–4014 (1995). ISSN 1079-7114. https://doi.org/10.1103/PhysRevLett.75.4011. http://www.ncbi.nlm.nih.gov/pubmed/10059792

36. A. Friedenauer, H. Schmitz, J.T. Glückert, D. Porras, T. Schaetz, Simulating a quantum magnet with trapped ions. Nat. Phys. **4**(10), 757–761 (2008). ISSN 1745-2473. http://dx.doi.org/10.1038/nphys1032

37. R. Blatt, D. Wineland, Entangled states of trapped atomic ions. Nature **453**(7198), 1008–1015 (2008). ISSN 0028-0836. https://doi.org/10.1038/nature07125. http://www.ncbi.nlm.nih.gov/pubmed/18563151. http://www.nature.com/articles/nature07125

38. H. Katori, S. Schlipf, H. Walther, Anomalous dynamics of a single ion in an optical lattice. Phys. Rev. Lett. **79**, 2221–2224 (1997). https://doi.org/10.1103/PhysRevLett.79.2221

39. C. Schneider, M. Enderlein, T. Huber, T. Schaetz, Optical trapping of an ion. Nat. Photonics **4**(11), 772–775 (2010). ISSN 1749-4885. http://dx.doi.org/10.1038/nphoton.2010.236

40. C. Schneider, M. Enderlein, T. Huber, S. Dürr, T. Schaetz, Influence of static electric fields on an optical ion trap. Phys. Rev. A **85**, 013422 (2012). https://doi.org/10.1103/PhysRevA.85.013422

41. C. Schneider, D. Porras, T. Schaetz, Experimental quantum simulations of many-body physics with trapped ions. Rep. Prog. Phys. **75**(2), 024401 (2012). http://stacks.iop.org/0034-4885/75/i=2/a=024401

42. M. Enderlein, T. Huber, C. Schneider, T. Schaetz, Single ions trapped in a one-dimensional optical lattice. Phys. Rev. Lett. **109**, 233004 (2012). https://doi.org/10.1103/PhysRevLett.109.233004

43. T. Huber, A. Lambrecht, J. Schmidt, L. Karpa, T. Schaetz, A far-off-resonance optical trap for a Ba$^+$ ion. Nat. Commun. **5**(5587) (2014)

44. A. Lambrecht, J. Schmidt, P. Weckesser, M. Debatin, L. Karpa, T. Schaetz, Long lifetimes and effective isolation of ions in optical and electrostatic traps. Nat. Photonics **11**(11), 704–707 (2017). ISSN 1749-4885. https://doi.org/10.1038/s41566-017-0030-2

45. J. Schmidt, A. Lambrecht, P. Weckesser, M. Debatin, L. Karpa, T. Schaetz, Optical trapping of ion coulomb crystals. Phys. Rev. X **8**(2), 021028 (2018). ISSN 2160-3308. https://doi.org/10. 1103/PhysRevX.8.021028. http://arxiv.org/abs/1712.08385

46. R.B. Linnet, I.D. Leroux, M. Marciante, A. Dantan, M. Drewsen, Pinning an ion with an intracavity optical lattice. Phys. Rev. Lett. **109**, 233005 (2012). https://doi.org/10.1103/ PhysRevLett.109.233005

47. L. Karpa, A. Bylinskii, D. Gangloff, M. Cetina, V. Vuletić, Suppression of ion transport due to long-lived subwavelength localization by an optical lattice. Phys. Rev. Lett. **111**, 163002 (2013). https://doi.org/10.1103/PhysRevLett.111.163002

48. T. Lauprêtre, R.B. Linnet, I.D. Leroux, H. Landa, A. Dantan, M. Drewsen, Controlling the potential landscape and normal modes of ion Coulomb crystals by a standing-wave optical potential. Phys. Rev. A **99**(3), 031401 (2019). ISSN 2469-9926. https://doi.org/10.1103/ PhysRevA.99.031401

49. Y. Huang, G.-Z. Li, S. Guan, A.G. Marshall, A combined linear ion trap for mass spectrometry. J. Am. Soc. Mass Spectrom. **8**(9), 962–969 (1997). ISSN 1879-1123. https://doi.org/10.1016/ S1044-0305(97)82945-5

50. T. Pruttivarasin, M. Ramm, I. Talukdar, A. Kreuter, H. Häffner, Trapped ions in optical lattices for probing oscillator chain models. New J. Phys. **13**(7), 075012 (2011). ISSN 1367-2630. https://doi.org/10.1088/1367-2630/13/7/075012. http://stacks.iop.org/1367-2630/13/i= 7/a=075012?key=crossref.37dab867a0b3575db6ed40692539a9a4

51. A. Benassi, A. Vanossi, E. Tosatti, Nanofriction in cold ion traps. Nat. Commun. **2**(1), 235–236 (2011). ISSN 20411723. https://doi.org/10.1038/ncomms1230

52. A. Bylinskii, D. Gangloff, V. Vuletic, Tuning friction atom-by-atom in an ion-crystal simulator. Science **348**(6239), 1115–1118 (2015). https://doi.org/10.1126/science.1261422

53. D. Gangloff, A. Bylinskii, I. Counts, W. Jhe, V. Vuletic, Velocity tuning of friction with two trapped atoms. Nat. Phys. **11**(11), 915–919 (2015). ISSN 1745-2473. http://dx.doi.org/10.1038/ nphys3459

54. J.D. Baltrusch, C. Cormick, G. De Chiara, T. Calarco, G. Morigi, Quantum superpositions of crystalline structures. Phys. Rev. A **84**, 063821 (2011). https://doi.org/10.1103/PhysRevA.84. 063821

55. J.D. Baltrusch, C. Cormick, G. Morigi, Quantum quenches of ion Coulomb crystals across structural instabilities. Phys. Rev. A **86**, 032104 (2012). https://doi.org/10.1103/PhysRevA.86. 032104

56. E. Shimshoni, G. Morigi, S. Fishman, Quantum zigzag transition in ion chains. Phys. Rev. Lett. **106**, 010401 (2011). https://doi.org/10.1103/PhysRevLett.106.010401

57. E. Shimshoni, G. Morigi, S. Fishman, Quantum structural phase transition in chains of interacting atoms. Phys. Rev. A **83**, 032308 (2011). https://doi.org/10.1103/PhysRevA.83. 032308

58. J.I. Cirac, P. Zoller, A scalable quantum computer with ions in an array of microtraps. Nature **404**(6778), 579–581 (2000). https://doi.org/10.1038/35007021

59. M. Pretko, L. Radzihovsky, Fracton-elasticity duality. Phys. Rev. Lett. **120**(19), 195301 (2018). ISSN 0031-9007. https://doi.org/10.1103/PhysRevLett.120.195301

60. T. Feldker, P. Bachor, M. Stappel, D. Kolbe, R. Gerritsma, J. Walz, F. Schmidt-Kaler, Rydberg excitation of a single trapped ion. Phys. Rev. Lett. **115**, 173001 (2015). https://doi.org/10.1103/ PhysRevLett.115.173001

61. G. Higgins, W. Li, F. Pokorny, C. Zhang, F. Kress, C. Maier, J. Haag, Q. Bodart, I. Lesanovsky, M. Hennrich, Single strontium Rydberg ion confined in a Paul trap. Phys. Rev. X **7**, 021038 (2017). https://doi.org/10.1103/PhysRevX.7.021038

62. G. Higgins, F. Pokorny, C. Zhang, Q. Bodart, M. Hennrich, Coherent control of a single trapped Rydberg ion. Phys. Rev. Lett. **119**, 220501 (2017). https://doi.org/10.1103/PhysRevLett.119. 220501

63. W.C. Campbell, J. Mizrahi, Q. Quraishi, C. Senko, D. Hayes, D. Hucul, D.N. Matsukevich, P. Maunz, C. Monroe, Ultrafast gates for single atomic qubits. Phys. Rev. Lett. **105**(9), 090502 (2010). ISSN 0031-9007. https://doi.org/10.1103/PhysRevLett.105.090502

Chapter 2
Trapping Ions with Light Fields

In principle, the optical forces experienced by ions are of the same origin as for the case of neutral atoms. This is due to the fact that the light shift responsible for the optical potential stems from the coupling of the light field to the outer electron of the atoms and not to the electric charge. Thus, in a field-free environment, the effective potentials generated by an optical field for an ion and a neutral atom of the same polarizability and energy level structure would be identical. So in order to obtain a trapping potential for an ion, we would only have to replace the potential generated by an external electric quadrupole field modulated at radiofrequency with suitable optical fields, e.g. a red-detuned Gaussian beam focused on the ion. In reality however, external electric fields are always finite, and in order to satisfy Laplace's equation,

$$\nabla^2 \Phi = 0 \tag{2.1}$$

where the quadratic potential of a quadrupole field is determined by $\Phi(x, y, z) \propto \alpha x^2 + \beta y^2 + \gamma z^2$ with the factors α, β, γ obeying the condition $\alpha + \beta + \gamma = 0$ [1], they cannot be arranged in a way to provide static confinement in all three dimensions. This is equivalent to a formulation found in Earnshaw's theorem: "An electrified body placed in a field of electric force cannot be in stable equilibrium" [2]. Laplace's equation imposes restrictions on the possible solutions Φ to the effect that no local extrema in free space are allowed, whereas local saddle points are possible. In other words, the curvature of a purely electrostatic potential is negative in at least one direction. To illustrate this aspect, let us consider an ion placed at the point of unstable equilibrium, i.e. maximum of a quadratic electric field with $\beta, \gamma > 0$ and $\alpha < 0$. This point coincides with the node of the radiofrequency field in a perfectly compensated Paul trap. For simplicity, let us also assume that the direction of the resulting deconfinement coincides with the x-axis as shown in Fig. 2.1. In presence of an optical trapping field originating from a Gaussian, red-detuned beam propagating along the z-axis, focused on the ion in a hypothetical

© The Author(s), under exclusive license to Springer Nature Switzerland AG 2019   9
L. Karpa, *Trapping Single Ions and Coulomb Crystals with Light Fields*,
SpringerBriefs in Physics, https://doi.org/10.1007/978-3-030-27716-1_2

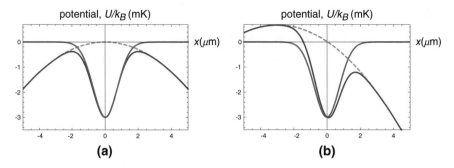

Fig. 2.1 (a) Effective potential U/k_B, with k_B the Boltzmann constant, (dark blue solid line) experienced by an ion in a repulsive quadratic electric field and an attractive optical potential: the effective trap depth is reduced as compared to the optical trap depth (red solid line) due to the defocusing electrostatic potential (gray dashed line) along the z-axis. (b) Accounting for a finite stray electric field leads to an additional tilt of the effective potential, further reducing its depth

scenario without any stray electric fields, an ion at $x = 0$ will be subject to a trapping optical potential with a reduced depth due to the negative curvature of the electric field. This fundamental reduction of the effective trap depth is drastically aggravated if an additional linear stray electric field is added, as illustrated on the right panel of Fig. 2.1.

While in realistic experiments the quadratic component of the electric field is determined by the configuration of electrostatic electrode parameters of the Paul trap, the linear contribution originates from uncompensated stray electric fields and is responsible for excess micromotion observed in all Paul traps. Already in the case of conventional ion traps great effort is invested in order to compensate such external fields and to minimize their detrimental impact on the performance of ion trap experiments.

In the case where we seek to replace the confinement provided by radiofrequency fields with optical forces, it is instructive to compare the available optical forces with those generated by uncontrollably deposited charges on the electrodes of a typical Paul trap. Assuming a corresponding potential of 1 V and ion to electrode distances of 1 mm , we obtain electric forces on the order of $F_{el,str} \sim 10^{-17}$ N, whereas with commercially available laser sources optical forces of about $F_{opt} \sim 10^{-19}$ N can be achieved. Under such conditions, Coulomb forces are orders of magnitude larger than optical forces, and trapping ions with optical fields would not be feasible. The conclusion of these considerations is that the obtainable performance of optical ion traps will hinge on the capability to detect and compensate stray electric fields as precisely as possible. Whether or not this can be done at a level sufficient for optical trapping was unclear until the first proof-of-principle demonstration of optical trapping of a single ion in 2010 [3]. The compensation accuracy achieved in this experiment, schematically shown in Fig. 2.2, was on the order of 100 mV/m, corresponding to residual electric forces of $F_{el,str} \sim 10^{-20}$ N.

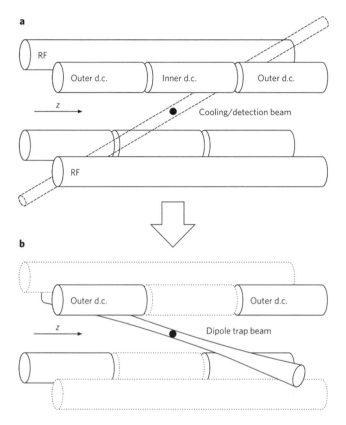

Fig. 2.2 Experimental setup used for the first realization of an optical trap for a single ion. (**a**) Initialization of a laser-cooled ion in a Paul trap with radial and axial confinement provided by rf electrodes and voltages applied to the dc electrodes, respectively. (**b**) During the optical trapping phase, the rf potentials are turned off. Radial confinement is now achieved with an optical dipole trap. Taken from [3]

In the following, we will discuss the prerequisites and experimental techniques for carrying out such experiments, followed by the description of a generic protocol for optical trapping.

2.1   Prerequisites for Optical Trapping Experiments

One of the basic prerequisites in all optical trapping experiments involving single atoms is a sufficient isolation from the environment. In contrast to optical tweezers experiments for small macroscopic objects, any collision with background gas leads to atom loss from the trap due to the comparatively small mass and typically low trap depths in such experiments. Most commonly, optical traps for atoms are realized under ultra-high vacuum conditions, with pressures of the background gas well below 10^{-9} mbar, resulting in collision rates below $1\,s^{-1}$ [4]. Analogous to the case of neutral atoms, in general the necessary preparation, manipulation, and detection rely on the availability of electronic transitions allowing for laser cooling and imaging. The latter is crucial for the stray electric field compensation schemes described in this work. This condition is met for a number of routinely employed ion species like Be^+, Mg^+, Ca^+, Zn^+, Sr^+, Cd^+, Ba^+, Yb^+, Lu^+, Hg^+ which have been laser cooled in Paul traps. While the capability to sympathetically cool ions that are not resonant with any of the present optical fields with high efficiency by embedding them in a laser cooled Coulomb crystal is an advantage of ions over neutral atom systems, this kind of experiments requires a substantially more refined level of control that has only been demonstrated recently [5].

As we will see in the following, the presence of a metastable electronic level can be used to obtain trapping potentials for ions that directly depend on the electronic state. This is the case not only for several commonly used elements, e.g. Yb^+, Ba^+, Ca^+, Sr^+ but also for more exotic species such as Lu^+ or Eu^+. The focus of this chapter will be on the simplest possible realizations of optical trapping of ions. A detailed discussion of the more advanced experiments with ion crystals will be presented later on. The fact that in the described experiments optical forces acting on the ions have to overcome their Coulomb interactions with ambient electric fields gives rise to a specific set of prerequisites.

As outlined in the previous section, the configuration of the electrostatic fields has a crucial impact on the effective trapping potential. In particular, any confining potential results in a negative curvature for at least one of the orthogonal directions, reducing the effective trap depth. Since it is advantageous to maintain a controlled static confinement, e.g. along the linear axis of commonly used Paul traps, a convenient starting point is to adjust the axial curvature as low as possible in order to stay within the stability region of the trap and to ensure efficient laser cooling. This can lead to secular frequencies of approximately $\omega_{ax} = 2\pi \times 10\,\text{kHz}$, which is at least an order of magnitude lower than the secular frequencies encountered in typical ion trapping experiments relying on a large separation of the vibrational levels, at least during the preparation phase. For such axial confinement settings, the negative curvature due to the electrostatic potential easily becomes comparable to the curvatures provided by even strongly focused high power lasers. In a second step, the radial compensation electrodes can be adjusted such that the resulting deconfinement is distributed equally among the two orthogonal radial directions. For a focused Gaussian beam propagating along the nodal line of the Paul trap this corresponds to a symmetric reduction of the effective radial potential and maximizes the trap depth for a given optical power in the dipole trap.

While in principle the detrimental consequences of the Coulomb interaction in an optical ion trapping experiment can be overcome by using sufficiently powerful laser sources, the vast majority of experiments largely benefits from a precise compensation of stray electric fields. It was shown that an accuracy level of approximately $100\,\text{mV/m}$ is sufficient for the proof-of-principle experiments carried out with commercially available laser sources [3, 6] and for certain applications such as controlled preparation of a predefined number of ions which will be discussed in the outlook section. On the other hand, applications in the field of ion–atom interactions require minute control over the effective potentials enabled by compensation uncertainties on the order of 10 mV/m and below [7].

As we will see in the following, the most straightforward way to fulfilling these requirements is to use a linear Paul trap in the outlined configuration for the preparation and detection steps. Paul traps with a large ion-electrode distance of several mm are particularly suited due to their compatibility with high numerical aperture optics crucial for precise stray field sensing and compensation, high resolution of the required electrostatic potentials, and intrinsically low anomalous heating rates. Nonetheless, most of the mentioned techniques are also applicable for planar traps, e.g. interfaced with buildup cavities as demonstrated in recent works utilizing a combination of radiofrequency and periodic optical fields [8–10].

2.2 Methodology

This section provides an overview over the most useful methods commonly employed for optical trapping of ions. These cover different aspects ranging from loading and suitable configurations of ion traps and preparation of the ions for an optical trapping sequence, to detection and analysis.

Paul Trap

A linear Paul trap with comparatively large ion to electrode distances of several millimeters such as the one depicted in Fig. 1.2b provides the following advantages in view of storing ions in dipole traps:

- Firstly, it provides a large trapping volume, allowing for efficient ion loading. Schemes involving photoionization with transitions in the ultraviolet spectral range, e.g. light at 405 nm used for the ionization of electronically excited Ba atoms, benefit from a reduced exposure time, mitigating the deposition of charges on dielectric surfaces such as windows or cavity mirrors. Continuous operation of an oven has been shown to produce increased drift rates for stray electric fields, such that ablative loading schemes bear the potential for improved performance of optical trapping experiments.
- Large numerical aperture optics allow for focusing of dipole trapping beams to small beam waists. For a given optical trap depth, the restoring force increases with smaller beam radii. Thus, smaller dipole traps provide more robustness against forces from stray charges. The same optical access can be used to collect ion fluorescence which, in combination with high spatial resolution of the ion's position, facilitates highly accurate stray field compensation [11, 12].

- High resolution of electrical potentials applied for compensation. With commercially available digital-to-analog converters, field resolutions on the order of 1 mV/m are readily achievable without the necessity for additional passive components such as voltage dividers which can alter the response of the system in a dynamical regime of operation.
- The so-called anomalous heating is the subject of active investigations, and has been shown to exhibit a strong dependence on the ion-to-electrode distance d, with a typical scaling of approximately d^{-4} [13, 14]. While in the case of planar traps with $d \sim 30\,\mu$m extensive measures have to be taken in order to ensure heating rates on the order of one vibrational quantum per ms [15], electrode-to-ion distances in the mm range provide several orders of magnitude lower heating rates. This is beneficial for example in the case of future investigations of ion–atom interactions in the quantum regime, and for potential applications in the field of quantum simulations with ions.

At the same time, large Paul traps invoke several disadvantages that can affect the performance of optical trapping experiments.

- A prominent example is the relatively low secular frequency of the radiofrequency confinement, making established methods for ground state cooling more challenging to implement.
- Maximized optical access can come at the expense of reduced shielding from ambient stray electric fields stemming from charges deposited on windows or other dielectric surfaces.
- The downside of a large trapping volume is its susceptibility to highly energetic ions stored on large trajectories within the trap that lead to heating and reduce the efficiency of experiments requiring a predefined number of ions due to random crystallization. The latter effect can be effectively mitigated once optical trapping conditions have been established, since the trapping region is then restricted to a much smaller volume given roughly by $V_{opt} \approx 2\pi w_0^2 z_R$, with w_0 being the waist radius and z_R the Rayleigh range of the dipole trapping beam.

Electrostatic Field Configuration

The effective potential for a single ion in a focused beam dipole trap in a direction $i \in \{x, y, z\}$ is a superposition of an electrostatic potential

$$U_{dc,i} = \frac{1}{2}m\omega_i^2 i^2 + eE_{str,i} i$$

comprising the attractive or repulsive contributions stemming from potentials applied to the Paul trap electrodes combined with the stray electric field $\vec{E}_{str}$, and the purely optical potential

$$U_{opt}(r, z) = U_0 \frac{\exp(-2r^2/w(z)^2)}{1 + (z/z_R)^2}$$

as illustrated in Fig. 2.1:

$$U_{tot} = U_{dc} + U_{opt}. \tag{2.2}$$

Here, $r, z, z_R, w(z)$ denote the radial and axial coordinates, the Rayleigh range, and beam radius, respectively. Therefore, the configuration of electrostatic fields is an important aspect that has to be taken into account in order to achieve optimal performance of optical traps. This configuration strongly depends on the parameters of the experiment and in particular on the chosen dipole trap geometry.

In order to mitigate the reduction of the effective potential depth along the directions of defocusing due to the applied electrostatic fields, it is helpful to use the following strategy: firstly, during the optical trapping phase the focusing potential contribution U_{dc} should be kept as weak as possible, and secondly, the resulting defocusing should be distributed equally along the two remaining orthogonal directions. This requires a method for measuring secular frequencies as well as a set of electrodes sufficient for tuning of the respective electrostatic potentials. Secular frequencies can be determined with an accuracy of at least 0.1 kHz by resonantly exciting laser cooled ions with modulated external electric potentials while monitoring their fluorescence and spatial extent e.g. with a charge-coupled device (CCD) camera. If the excitation occurs at the secular frequency of the ion, a distinct expansion can be observed. In recent experiments with focused beam dipole traps, typical values between $\omega_{dc} = 2\pi \times 10$ kHz and $2\pi \times 20$ kHz have proven to be a suitable choice.

In future experiments aiming to use the technique of optical trapping for confining 2d ion Coulomb crystals a different trapping geometry is advantageous. A straightforward looking way to approach this problem would be to increase the axial confinement of an initially linear chain to a point where a structural transition to a zig-zag configuration and later on to a 2d crystal takes place while simultaneously increasing the optical intensity of the dipole trap to compensate the resulting defocusing from the electrostatic field and mutual Coulomb repulsion between the ions. This approach invokes the disadvantage of having to use comparatively large beam waists to fit in the Coulomb crystals and correspondingly very large optical powers that easily can be in excess of commercially available solutions. Instead, one might use a standing wave geometry to provide the axial confinement, and use electrostatic potentials for radial confinement. This arrangement should provide both sufficient robustness with respect to axial defocusing and stray electric fields due to a much stronger gradient of the optical potential achievable in optical potentials with a λ-scale periodicity. This aspect is also discussed in Sect. 5.2.

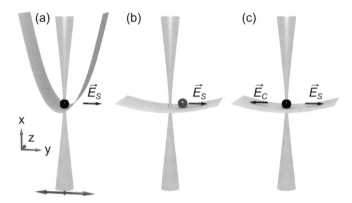

Fig. 2.3 Principle of the a.c. Stark shift-assisted stray electric field compensation scheme. A residual stray electric field E_S causes a displacement of the ion in (**a**) a harmonic rf pseudopotential with stiff confinement (gray parabolic surface) and shallow confinement (**b, c**). The initial displacement is amplified by the factor $\left(\omega_{rf,h}/\omega_{rf,l}\right)^2$. Compensation potentials are then used to minimize the displacement resulting in a residual uncertainty of $\delta E_S < 9\,\text{mV/m}$. Taken from [12]

Stray Electric Field Compensation

Since the rf fields are switched off during the optical trapping period, the ion does not experience micromotion. On the other hand, the same methods applied for minimizing excess micromotion are also effective in reducing the influence of stray electric fields. While several methods have been developed over the last decades, in the described setting an approach based on measuring the displacement of an ion exposed to a static stray field at different rf confinements with secular frequencies $\omega_{rf,h}$ and $\omega_{rf,l}$, as illustrated in Fig. 2.3, has proven particularly sensitive. This method is an adapted version of ideas described in 1998 by Berkeland and colleagues [11]: "In the first of these methods, which is sensitive to excess micromotion caused by static fields, the time-averaged ion position is monitored as the pseudopotential is raised and lowered".

In this scheme, the achievable stray field detection accuracy scales as the square of the ratio of the secular frequencies, i.e. $\delta E_S \propto \left(\frac{\omega_{rf,l}}{\omega_{rf,h}}\right)^2$, such that the capability to switch between a stiff and a shallow rf-potential, e.g. with $\omega_{rf,h} \approx 2\pi\,300\,\text{kHz}$ and $\omega_{rf,l} \approx 2\pi\,30\,\text{kHz}$, is highly advantageous. Using the light shift induced by the focused dipole trap beam as a marker for the node of the rf field and as a way to derive position when moving an ion through the beam, a residual uncertainty of $9\,\text{mV/m}$ has been demonstrated, with the corresponding measurements being carried out within about 90 s [12]. Apparatuses with sufficiently high spatial resolution of the imaging system in the radial directions benefit from the capability to directly monitor the ion position for different rf confinement settings. This scheme yields an accuracy between 5 and $10\,\text{mV/m}$ after a measurement time of about 30 s with a position resolution of 200 nm.

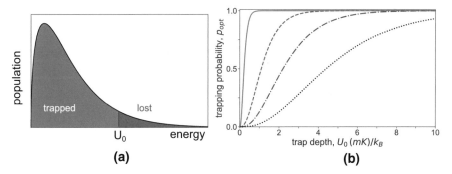

Fig. 2.4 Illustration of expected trapping probability p_opt in a trap of depth U_0 according to a simplified model considering a truncated Maxwell–Boltzmann distribution. (**a**) The ratio of the accumulated populations in all energy levels below U_0 and above the threshold determines the expected probability for a successful experimental realization. (**b**) Examples of expected trapping probabilities based on the radial cutoff model for ions prepared at initial temperatures of 0.1 mK (solid line), 0.5 mK (dashed line), 1.0 mK (dash-dotted line), and 2.0 mK (dotted line) as a function of applied trap depth U_0

Temperature Measurement

The trapping probability of an atom in a potential with depth U_0 depends on the initial temperature of the atom T as well as on the trap depth. In the simplified model based on a truncated Boltzmann distribution as depicted in Fig. 2.4a the probability p_{en} for a successful experimental realization is given by the following expression [16]:

$$p_{en}(\xi) = 1 - \left(\frac{\xi^2}{2} + \xi + 1\right) e^{-\xi},$$

with $\xi = \frac{U_0}{k_B T}$. This simplified picture neglecting energy stored in the angular momentum of the atoms can be extended to a more complete radial cutoff model which yields a modified expression for the survival probability [17]:

$$p_{rad}(\xi) = 1 - e^{-2\xi} - 2\xi e^{-\xi}. \tag{2.3}$$

According to this result, the temperature of an atom can be determined by measuring the survival probability for different trap depths as illustrated in Fig. 2.4b. This method provides the advantage of measuring the temperature for a wide range of trapping durations τ directly within the trap without additional manipulation of the atoms. However, since the actual energy of the atom can be fixed during the trapping phase and the outcome of a trapping attempt is binary, this approach requires accumulation of statistics by repeating the experiment several times in order to obtain a sufficiently accurate representation of the energy distribution.

Optical Trapping Protocol

A generic sequence suitable for optical trapping of ions, schematically represented in Fig. 2.5, comprises the following steps subdivided in three phases:

- Phase 1: preparation

  - loading into Paul trap
  - Doppler cooling
  - stray field compensation

- Phase 2: transfer and trapping

  - ramp optical dipole trap to U_0
  - ramp rf field amplitude to zero (not dc potentials)
  - trap ion optically for a duration τ

- Phase 3: detection

  - switch Paul trap on
  - switch dipole trap off
  - fluorescence detection

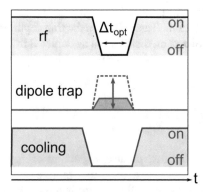

Fig. 2.5 The experimental protocol comprises three phases: (1) Preparation: loading and Doppler cooling the ions, stray field compensation; (2) Trapping: transfer into the dipole trap by turning off the rf field and cooling lasers for the optical trapping duration; (3) Detection: ion fluorescence is used to determine if the trapping attempt was successful

The phases 1 and 3 of the optical trapping sequence are straightforward in view of the previously described methods. The transfer between the rf and optical traps however requires taking certain precautions, since the Paul trap is operated outside of the optimized settings within the stability diagram and ultimately is rendered unstable, at which point the optical confinement has to take over. A critical parameter is the ramp duration of the rf potential and the dipole trap, respectively.

It has to be chosen carefully in order to maintain adiabaticity while applying the changes fast enough to mitigate the detrimental effects of crossing instabilities encountered in rf traps with higher order anharmonicities [18]. At radial secular frequencies on the order of $\omega_{rf,rad} \approx 2\pi \times 100$ kHz typical ramp durations of about $50\,\mu s$ have been shown to provide low additional heating and reliable results.

Optical Trapping Probabilities

Optical ion trapping attempts are realizations of Bernoulli experiments and as such yield inherently binary results. With the detection being based on observing the fluorescence of an ion after a trapping attempt, the achievable fidelities can approach 100%. The ratio of successful and total attempts N_S/N_T can be used to calculate the most likely underlying probability. The uncertainties of such measurements can then be estimated by calculating corresponding Wilson score intervals [19] and are purely statistical. The exact value of the uncertainties depends on N_S/N_T, but typically, in order to obtain uncertainties of about $\pm 15\%$, approximately 20 attempts are required.

References

1. D. Leibfried, R. Blatt, C. Monroe, D. Wineland, Quantum dynamics of single trapped ions. Rev. Mod. Phys. **75**, 281–324 (2003). https://doi.org/10.1103/RevModPhys.75.281
2. J.C. Maxwell, *A Treatise on Electricity and Magnetism*, vol. 1 (Oxford, Clarendon Press, 1873)
3. Ch. Schneider, M. Enderlein, T. Huber, T. Schaetz, Optical trapping of an ion. Nat. Photonics **4**(11), 772–775 (2010). ISSN 1749-4885. http://dx.doi.org/10.1038/nphoton.2010.236
4. R. Grimm, M. Weidemüller, Y.B. Ovchinnikov, Optical dipole traps for neutral atoms. Adv. At. Mol. Opt. Phys. **42**, 95–170 (2000)
5. J. Schmidt, A. Lambrecht, P. Weckesser, M. Debatin, L. Karpa, T. Schaetz, Optical trapping of ion coulomb crystals. Phys. Rev. X **8**(2), 021028 (2018). ISSN 2160-3308. https://doi.org/10.1103/PhysRevX.8.021028. http://arxiv.org/abs/1712.08385
6. M. Enderlein, T. Huber, C. Schneider, T. Schaetz, Single ions trapped in a one-dimensional optical lattice. Phys. Rev. Lett. **109**, 233004 (2012). https://doi.org/10.1103/PhysRevLett.109.233004
7. M. Cetina, A.T. Grier, V. Vuletić, Micromotion-induced limit to atom-ion sympathetic cooling in Paul traps. Phys. Rev. Lett. **109**, 253201 (2012). https://doi.org/10.1103/PhysRevLett.109.253201
8. L. Karpa, A. Bylinskii, D. Gangloff, M. Cetina, V. Vuletić, Suppression of ion transport due to long-lived subwavelength localization by an optical lattice. Phys. Rev. Lett. **111**, 163002 (2013). https://doi.org/10.1103/PhysRevLett.111.163002
9. A. Bylinskii, D. Gangloff, V. Vuletic, Tuning friction atom-by-atom in an ion-crystal simulator. Science **348**(6239), 1115–1118 (2015). https://doi.org/10.1126/science.1261422
10. D. Gangloff, A. Bylinskii, I. Counts, W. Jhe, V. Vuletic, Velocity tuning of friction with two trapped atoms. Nat. Phys. **11**(11), 915–919 (2015). ISSN 1745-2473. http://dx.doi.org/10.1038/nphys3459

11. D.J. Berkeland, J.D. Miller, J.C. Bergquist, W.M. Itano, D.J. Wineland, Minimization of ion micromotion in a Paul trap. J. Appl. Phys. **83**, 10 (1998)
12. T. Huber, A. Lambrecht, J. Schmidt, L. Karpa, T. Schaetz, A far-off-resonance optical trap for a Ba^+ ion. Nat. Commun. **5**, 5587 (2014)
13. L. Deslauriers, S. Olmschenk, D. Stick, W.K. Hensinger, J. Sterk, C. Monroe, Scaling and suppression of anomalous heating in ion traps. Phys. Rev. Lett. **97**, 103007 (2006). https://doi.org/10.1103/PhysRevLett.97.103007
14. Q.A. Turchette, D. Kielpinski, B.E. King, D. Leibfried, D.M. Meekhof, C.J. Myatt, M.A. Rowe, C.A. Sackett, C.S. Wood, W.M. Itano, C. Monroe, D.J. Wineland, Heating of trapped ions from the quantum ground state. Phys. Rev. A **61**(6), 063418 (2000). ISSN 1050-2947. https://doi.org/10.1103/PhysRevA.61.063418
15. D.A. Hite, Y. Colombe, A.C. Wilson, K.R. Brown, U. Warring, R. Jördens, J.D. Jost, K.S. McKay, D.P. Pappas, D. Leibfried, D.J. Wineland, 100-fold reduction of electric-field noise in an ion trap cleaned with in situ argon-ion-beam bombardment. Phys. Rev. Lett. **109**(10), 103001 (2012). ISSN 0031-9007. https://doi.org/10.1103/PhysRevLett.109.103001
16. C. Tuchendler, A.M. Lance, A. Browaeys, Y.R.P. Sortais, P. Grangier, Energy distribution and cooling of a single atom in an optical tweezer. Phys. Rev. A **78**, 033425 (2008)
17. C. Schneider, M. Enderlein, T. Huber, S. Dürr, T. Schaetz, Influence of static electric fields on an optical ion trap. Phys. Rev. A **85**, 013422 (2012). https://doi.org/10.1103/PhysRevA.85.013422
18. R. Alheit, C. Hennig, R. Morgenstern, F. Vedel, G. Werth, Observation of instabilities in a Paul trap with higher-order anharmonicities. Appl. Phys. B **61**(3), 277–283 (1995). ISSN 0946-2171. https://doi.org/10.1007/BF01082047
19. E.B. Wilson, Probable Inference, the Law of Succession, and Statistical Inference. J. Am. Stat. Assoc. **22**(158), 209–212 (1927). https://doi.org/10.1080/01621459.1927.10502953

Chapter 3
Optical Dipole Traps for Single Ions

The first demonstration of optical trapping of an ion was achieved in 2010, where a near resonant focused dipole trap was used to confine a Mg^+ ion for a few milliseconds [1]. As discussed in the previous chapters, already at this early stage fairly precise compensation of stray electric fields was essential. In order to obtain trap depths of several $(10\,mK)k_B$, sufficient for optical trapping of ions laser cooled to about $5\,T_D$, a focused beam with a detuning of about $\Delta \approx 6.6 \times 10^3\,\Gamma$ was used, where Γ is the natural line width of the addressed optical transition and the Doppler cooling limit is denoted with $T_D \approx 1\,mK$. It was found that the heating caused by off-resonant scattering from the dipole trap experienced by the ion within trapping durations of a few ms was on the order of the trap depth and thus the main mechanism limiting the observed ion lifetime as shown in Fig. 3.1.

Later on, this experiment was successfully carried out within a standing wave of the same wavelength [2]. The lifetimes in this experiment were limited to approximately $100\,\mu s$, due to the increased off-resonant scattering in the standing wave and due to limitations during the transfer between the rf and optical traps.

A similar approach was used in a different set of experiments combining radial rf confinement with an optical standing wave along the third direction. There, the influence of the optical potentials was also demonstrated by observing pinning [3] or strongly suppressed ion transport [4]. In the latter work it was also shown that the same techniques can be extended to confine small chains with up to three ions for at least 10 ms, while simultaneously cooling them close to the ground state and monitoring their individual positions with sub-wavelength resolution. The lifetime of the ions in the 1d standing wave potential was mainly limited by externally modulated electric fields causing additional heating. An extrapolation to the unperturbed case where the ions are left in the dark state shows that the expected lifetime can reach several seconds, unless other fundamental limitations inhibit trapping on these timescales.

© The Author(s), under exclusive license to Springer Nature Switzerland AG 2019
L. Karpa, *Trapping Single Ions and Coulomb Crystals with Light Fields*,
SpringerBriefs in Physics, https://doi.org/10.1007/978-3-030-27716-1_3

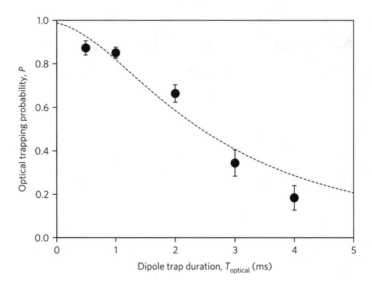

Fig. 3.1 Optical trapping probability of a ^{24}Mg$^+$ ion in a near-resonant optical dipole trap as a function of the trapping duration. Taken from [1]

A very useful property of dipole traps for the simplified case of a two-level atom is the scaling of the potential depth as:

$$U_0(r) = I(r) \left(\frac{\Gamma}{\Delta} \right) \frac{3\pi c^2}{2\omega_0^3}, \tag{3.1}$$

where c is the speed of light, ω_0 the resonance frequency of the electronic transition, and I is the intensity of the beam at the location of the atoms, while the off-resonant scattering rate Γ_{sc} is given by:

$$\Gamma_{sc}(r) = I(r) \left(\frac{\Gamma}{\Delta} \right)^2 \frac{3\pi c^2}{2\hbar\omega_0^3} \tag{3.2}$$

[5]. Thus, for a constant trap depth, the influence of off-resonant scattering is suppressed by trapping atoms in far-detuned dipole traps operated at high intensities.

A derivation of the Eqs. 3.1 and 3.2 can be found in contemporary textbooks on atomic physics, e.g. [6], and in [5]. It should be noted that these relations hold for the case of a two-level atom interacting with far-detuned laser radiation such that the rotating wave approximation is valid. In practice, the trapping light fields interact with multilevel atoms such that the contributions from different electronic states have to be taken into account. For instance, the presence of degenerate magnetic levels of the D-manifolds of Ba$^+$ can lead to a strongly reduced scattering rate on the repumping transition [7] due to population of dark states. By extension, the trapping potential of ions shelved into one of the D-manifolds is expected to exhibit a strong dependence on the composition of the final state. However, the conclusion that the off-resonant scattering rate can be reduced by increasing the detuning still holds in the generalized case, since the effective light shift is essentially a sum of the individual energy shifts from all of the coupled electronic states after accounting for the respective detunings and line strengths. Accordingly, the Eqs. 3.1 and 3.2 are applicable to each of these individual transitions.

3.1   Far-Off-Resonance Optical Traps for Ions

Following this approach, single Ba^+ ions were optically trapped in a far-off-resonance optical trap (FORT) at a wavelength of 532 nm [8]. The drastically increased detuning compared to [1] resulted in a reduction of the off-resonant scattering rate by three orders of magnitude and the corresponding recoil heating rate by four orders of magnitude. In comparison to the previous proof-of-principle experiments carried out with Mg^+, the barium ions exhibit a richer energy level structure. In particular, the presence of metastable $5 D_{3/2}$ and $5 D_{5/2}$ levels as illustrated in Fig. 3.2b allows to implement state-selective potentials. In the specific configuration used in the first experiments, all ions in the said D levels experienced a repulsive potential leading to ion loss as a consequence of the dipole trap being blue-detuned to the $5 D_{3/2} \longrightarrow 6 P_{1/2}$ and $5 D_{5/2} \longrightarrow 6 P_{3/2}$ transitions. The timescale for the population of these repulsively interacting states was given by the off-resonant scattering rate, resulting in lifetimes on the order of a few ms. Together with the improved accuracy in the detection of stray electric fields, this performance was already of interest for future investigations in the field of ion–atom interaction in a rf-free environment.

However, with the potential complications arising in the regime of low temperatures and high atomic densities for certain combinations of ions and atoms, such as $^{138}Ba^+$ and ^{87}Rb, an increased lifetime can afford additional flexibility in the choice of initial parameters. For example, the increased occurrence of three-body recombination events in presence of an ion impurity [10] can make it advantageous to operate in a regime of low density below $10^{12}\,cm^{-3}$, where the strong reduction of elastic collision processes requires a longer interaction duration. An interesting question in view of such research perspectives is what lifetimes realistically can be achieved. A series of experiments addressing this issue has recently been carried out using an experimental setup suitable for such investigations, as depicted in Fig. 3.2a.

As outlined in Chap. 2, Sect. 2.2 an increased performance of optical traps can be achieved by using a Paul trap with high numerical aperture access to the ion as well as by improving the accuracy of stray electric field compensation and cooling during the preparation phase. Focusing dipole traps to smaller radii increases the available restoring force for a given intensity, enhancing the effective trap depth. According to Eq. 2.3 the trapping probability scales exponentially with the ratio $\xi^{-1} = k_B T / U_0$ such that a lower initial temperature leads to a significantly higher trapping probability for a given dipole trap depth, as illustrated in Fig. 2.4b. The

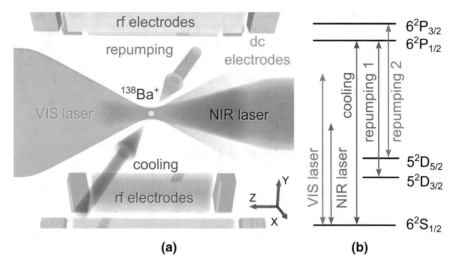

Fig. 3.2 (**a**) Schematic setup for optical trapping of ^{138}Ba$^+$ in dipole traps operated at $\lambda_{VIS} = 532$ nm (VIS) and $\lambda_{NIR} = 1064$ nm (NIR). In addition to lasers used for Doppler cooling (blue arrow) the ion was exposed to repumping lasers (short orange arrow) for depopulating metastable D level manifolds during the trapping duration. (**b**) Electronic levels and optical transitions used for Doppler cooling ($\lambda_{Doppler} = 493$ nm), repumping ($\lambda_{rp1} = 650$ nm, $\lambda_{rp2} = 615$ nm), and optical trapping. Adapted from [9]

combination of these techniques allows to use comparably low intensities at a constant level of performance, effectively reducing the off-resonant scattering rate and hence ion loss due to population of the D states.

Implementing these improvements as compared to the first experiment demonstrating a FORT for Ba$^+$ ions [8], in particular reducing the beam waist radius from 3.9 to 2.6 μm and lowering the temperature after laser cooling from 8.5 mK to the Doppler limit of 365 μK, the $1/e$ lifetime of an ion in an optical trap was significantly extended to ∼20 ms, as shown in Fig. 3.3. It has been shown that population in the D levels originating from off-resonant scattering can be transferred back to the initial $5S_{1/2}$ state by illuminating the ion with additional repumping lasers during the optical trapping phase. Consequently, the corresponding reduction of ion loss is reflected in a sizeable enhancement of the lifetime by about an order of magnitude, in comparison to the case where no repumping is applied, to about 170 μs. It should be noted that the branching ratio of the $P \longrightarrow S$ and the $P \longrightarrow D$ transitions is about 3 : 1, which is quite unfavorable in comparison with other commonly employed species such as Yb$^+$, Sr$^+$, Ca$^+$. The same experiment carried out, e.g. with ^{171}Yb$^+$ featuring a branching ratio from $2P_{1/2}$ to $2D_{3/2}$ state of ∼0.005, is expected to show a suppression of the loss probability via the metastable state by a factor of approximately 67 per scattered photon. Elements such as Be$^+$, Mg$^+$, Zn$^+$, Cd$^+$, and Hg$^+$ provide a closed cycling transition, such that off-resonant scattering can lead to recoil heating but not to the population of repulsively interacting electronic states and thus to loss.

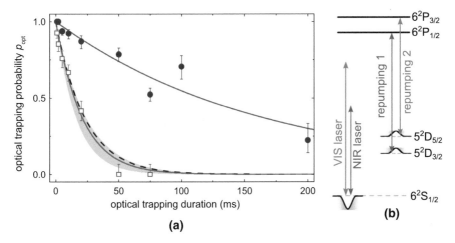

Fig. 3.3 (**a**) Dependence of p_{opt} on trapping duration Δt_{opt}. Open squares: data taken with the dipole laser only. Dashed line: expected lifetime of $\tau_D = 23 \pm 4$ ms based on measured off-resonant scattering rates into the D manifolds. Shaded area: bounds of a theoretical prediction $\tau_{th} = 20 \pm 4$ ms. Circles: data taken with additional repumping lasers. Solid lines: exponential fits to each data set, showing an increase of the lifetime in the dipole trap at 532 nm, $\tau_{VIS} = 21 \pm 2$ ms to $\tau_{VIS,rp} = 166 \pm 19$ ms due to repumping. (**b**) Schematic of the relevant electronic states coupled by the dipole traps and the repumping lasers. The light shifts induced by the VIS dipole trap in the S state (lowering the unperturbed energy) and in the D states (increasing the energy). Adapted from [9]

In view of potential applications in the realm of quantum information processing or quantum simulations, it is important to note that the coherence time in the dipole trap τ_{coh} is not prolonged by repumping. In fact, any coherent evolution is expected to be disrupted with the occurrence of the first off-resonant scattering event. A viable strategy for increasing τ_{coh} would be to lower the initial temperature by implementing standard ground state cooling techniques, which in principle are compatible with optical potentials [4, 11]. Making use of blue-detuned standing waves may also significantly reduce off-resonant scattering by confining the ions in the nodes of the optical field [4]. An alternative route would be to implement even further detuned dipole traps, or ultimately quasi-electrostatic traps (QUEST) with scattering rates below 10^{-3} s^{-1}, for example by using CO_2 laser sources [5].

3.2 Lifetime of an Ion in a Further Detuned Optical Trap

As shown in the previous chapter, the main limitation with respect to lifetimes of Ba$^+$ ions in optical traps is related to the off-resonant scattering rate Γ_{offr}. An effective and straightforward approach to reducing Γ_{offr} has been laid out decades ago in experiments with neutral atoms. As implied by Eqs. 3.1 and 3.2,

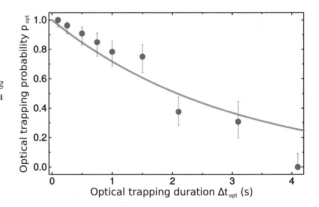

Fig. 3.4 Ion lifetime in the NIR trap. Dependence of optical trapping probability p_{opt} on trapping duration Δt_{opt} for single Ba$^+$ ions. Line: result of a fit, assuming exponential decay, yielding a lifetime of $\tau_{NIR} = (3 \pm 0.3)$ s. Taken from [9]

it is advantageous to increase the detuning of the dipole trap, while increasing its intensity in order to maintain a constant trap depth. In the described experiment, this effect can be directly observed under identical conditions in the same apparatus by comparing the lifetime observed in a VIS dipole trap at 532 nm with results obtained in a NIR dipole trap at 1064 nm. As shown in Fig. 3.4, the change in detuning from $\Delta_{VIS} = 3 \times 10^6\,\Gamma$ to $\Delta_{NIR} = 2 \times 10^7\,\Gamma$ leads to a dramatic increase of the lifetime to $\tau_{NIR} = 3 \pm 0.3$ s, now reaching trapping durations on the order of seconds. The observed lifetimes are comparable with the results achieved in experiments carried out in a comparable regime with single neutral atoms in dipole traps [11–14]. This may indicate that the lifetime could be limited by collisions with background gas, rather than due to fundamental effects or other mechanisms specific to the investigated system. For example, intensity fluctuations of the dipole trap lead not only to heating as is the case in neutral atom systems, but also to a displacement of the ion. Similarly, fluctuations of ambient fields result in random changes of the ion's position within the dipole trap, accompanied by correlated intensity fluctuations.

3.3 Heating Rate Due to Ambient Fields

In order to investigate the influence of the environment including the coupling of the trapped atom to electric fields, the optical trapping sequence was adapted by including an additional waiting time of 500 ms, allowing to use the temperature measurement method described in Sect. 2.2 to determine a heating rate. During this additional period, the ion was exposed to the maximum available intensity of the dipole trap, as well as to a residual rf field as indicated in the inset of Fig. 3.5. The result of such a measurement shown in Fig. 3.5 therefore provides an upper bound for the heating rate of $R_{max} = 400\,\mu\mathrm{K\,s}^{-1}$. Under realistic conditions, even at the Doppler limit much lower intensities can be used to confine the ion and can be reduced further by implementing sub-Doppler cooling techniques. In

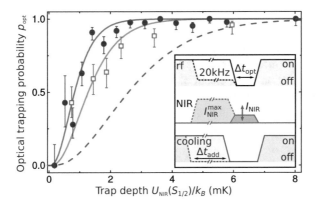

Fig. 3.5 Heating rate measurement in the dipole trap. p_{opt} is measured for optical trapping directly after Doppler cooling of the ion (blue circles) and after a delay of $\Delta t_{add} = 500$ ms, including exposure of the ion to an optical field with maximal intensity $I_{NIR,max}$, ambient and weak radiofrequency fields (red squares). During Δt_{add} laser cooling was turned off. The temperature was probed for $\Delta t_{opt} = 10$ ms by following the sequence depicted in the inset. Solid lines: fits to the data assuming a radial cutoff model, with temperatures $T_1 = 320 \pm 30\,\mu$K and $T_2 = 500 \pm 60\,\mu$K. Taken from [9]

addition, a substantial reduction of the heating rate can be expected by implementing the following optimizations not adapted in the conceptual investigations described above: stabilization of laser intensity and beam pointing, optimization of electric noise filtering, and improved shielding from ambient fields by adapting the trap geometry.

It should also be noted that in envisioned ion–atom collision experiments, which is one of the highlighted scenarios that could strongly benefit from absence of rf fields and the related micromotion-induced heating, elastic collisions are expected to be a highly efficient cooling mechanism [15, 16] with cooling rates that are several orders of magnitude higher than R_{max}. For the case of Ba$^+$ ions immersed in cloud of ultracold Rb atoms, the sympathetic cooling rate is estimated to be on the order of 100 mK s^{-1} [9].

In view of potential applications in the field of quantum simulations and information processing it is instructive to compare these findings with typical heating rates obtained in planar traps as one of the most promising currently available platforms in this field of research. While it is known that proximity to electrode surfaces can result in the so-called anomalous heating [17, 18], recent developments show that this effect can be reduced by one or more than two orders of magnitude by cooling the electrodes [18] or by employing techniques such as argon-ion-beam bombardment [19], respectively. For example, it was shown that the heating rate for a Be$^+$ ion positioned $40\,\mu$m above the chip surface can be lowered from initially $\sim$7000 quanta s^{-1} to $\sim$40 quanta s^{-1} for a secular frequency of $(2\pi)3.6$ MHz or about 7 mK s^{-1} [19]. In comparison, the upper bound for the heating rate measured in the optical trap is substantially lower, such that optical traps for ions may be a promising complementary approach to other currently pursued techniques. However, the coupling of electric field noise to ions confined in an optical potential is expected to be non-trivial and at this point it is not clear what effects can be expected when optical traps are operated close to or in combination with electrode structures. Therefore, further experimental investigations of decoherence and heating mechanisms in optical dipole traps for ions are indispensable in order to quantitatively gauge the potential gain by applying these novel methods to QIP and quantum simulations.

References

1. C. Schneider, M. Enderlein, T. Huber, T. Schaetz, Optical trapping of an ion. Nat. Photonics **4**(11), 772–775 (2010). ISSN 1749-4885. http://dx.doi.org/10.1038/nphoton.2010.236
2. M. Enderlein, T. Huber, C. Schneider, T. Schaetz, Single ions trapped in a one-dimensional optical lattice. Phys. Rev. Lett. **109**, 233004 (2012). https://doi.org/10.1103/PhysRevLett.109.233004
3. R.B. Linnet, I.D. Leroux, M. Marciante, A. Dantan, M. Drewsen, Pinning an ion with an intracavity optical lattice. Phys. Rev. Lett. **109**, 233005 (2012). https://doi.org/10.1103/PhysRevLett.109.233005.
4. L. Karpa, A. Bylinskii, D. Gangloff, M. Cetina, V. Vuletić. Suppression of ion transport due to long-lived subwavelength localization by an optical lattice. Phys. Rev. Lett. **111**, 163002 (2013). https://doi.org/10.1103/PhysRevLett.111.163002
5. R. Grimm, M. Weidemüller, Y.B. Ovchinnikov, Optical dipole traps for neutral atoms. Adv. At. Mol. Opt. Phys. **42**, 95–170 (2000)
6. C.J. Foot, *Atomic Physics*. Oxford Master Series in Physics (Oxford University Press, Oxford, 2005). ISBN 9780198506966. https://books.google.de/books?id=kXYpAQAAMAAJ
7. D.J. Berkeland, M.G. Boshier, Destabilization of dark states and optical spectroscopy in Zeeman-degenerate atomic systems. Phys. Rev. A **65**, 033413 (2002). https://doi.org/10.1103/PhysRevA.65.033413
8. T. Huber, A. Lambrecht, J. Schmidt, L. Karpa, T. Schaetz, A far-off-resonance optical trap for a Ba$^+$ ion. Nat. Commun. **5**, 5587 (2014)
9. A. Lambrecht, J. Schmidt, P. Weckesser, M. Debatin, L. Karpa, T. Schaetz, Long lifetimes and effective isolation of ions in optical and electrostatic traps. Nat. Photonics **11**(11), 704–707 (2017). ISSN 1749-4885. https://doi.org/10.1038/s41566-017-0030-2
10. A. Härter, A. Krükow, A. Brunner, W. Schnitzler, S. Schmid, J.H. Denschlag, Single ion as a three-body reaction center in an ultracold atomic gas. Phys. Rev. Lett. **109**, 123201 (2012). https://doi.org/10.1103/PhysRevLett.109.123201
11. A.M. Kaufman, B.J. Lester, C.A. Regal, Cooling a single atom in an optical tweezer to its quantum ground state. Phys. Rev. X **2**(4), 1–7 (2012)
12. D. Barredo, S. de Léséleuc, V. Lienhard, T. Lahaye, A. Browaeys, An atom-by-atom assembler of defect-free arbitrary two-dimensional atomic arrays. Science **354**(6315), 1021–1023 (2016)
13. M. Endres, H. Bernien, A. Keesling, H. Levine, E.R. Anschuetz, A. Krajenbrink, C. Senko, V. Vuletic, M. Greiner, M.D. Lukin, Atom-by-atom assembly of defect-free one-dimensional cold atom arrays. Science **354**(6315), 1024–1027 (2016). ISSN 0036-8075. https://doi.org/10.1126/science.aah3752. http://www.sciencemag.org/lookup/doi/10.1126/science.aah3752
14. T. Xia, M. Lichtman, K. Maller, A.W. Carr, M.J. Piotrowicz, L. Isenhower, M. Saffman, Randomized benchmarking of single-qubit gates in a 2D array of neutral-atom qubits. Phys. Rev. Lett. **114**(10), 100503 (2015)

15. M. Tomza, C.P. Koch, R. Moszynski, Cold interactions between an Yb^+ ion and a Li atom: prospects for sympathetic cooling, radiative association, and Feshbach resonances. Phys. Rev. A **91**(4), 042706 (2015). https://doi.org/10.1103/physreva.91.042706

16. M. Krych, W. Skomorowski, F. Pawłowski, R. Moszynski, Z. Idziaszek, Sympathetic cooling of the Ba^+ ion by collisions with ultracold Rb atoms: theoretical prospects. Phys. Rev. A **83**, 032723 (2011). https://doi.org/10.1103/PhysRevA.83.032723

17. Q.A. Turchette, D. Kielpinski, B.E. King, D. Leibfried, D.M. Meekhof, C.J. Myatt, M.A. Rowe, C.A. Sackett, C.S. Wood, W.M. Itano, C. Monroe, D.J. Wineland, Heating of trapped ions from the quantum ground state. Phys. Rev. A **61**(6), 063418 (2000). ISSN 1050-2947. https://doi.org/10.1103/PhysRevA.61.063418

18. L. Deslauriers, S. Olmschenk, D. Stick, W.K. Hensinger, J. Sterk, C. Monroe, Scaling and suppression of anomalous heating in ion traps. Phys. Rev. Lett. **97**, 103007 (2006). https://doi.org/10.1103/PhysRevLett.97.103007

19. D.A. Hite, Y. Colombe, A.C. Wilson, K.R. Brown, U. Warring, R. Jördens, J.D. Jost, K.S. McKay, D.P. Pappas, D. Leibfried, D.J. Wineland, 100-fold reduction of electric-field noise in an ion trap cleaned with in situ argon-ion-beam bombardment. Phys. Rev. Lett. **109**(10), 103001 (2012). ISSN 0031-9007. https://doi.org/10.1103/PhysRevLett.109.103001

Chapter 4
Optical Trapping of Coulomb Crystals

So far, optical traps in the previously described experiments have been used to trap a single ion. As was shown in the previous chapters, the techniques adapted from neutral atom traps to ions are straightforwardly applicable, if the additional contributions to the effective potential originating from electrostatic defocusing (see Fig. 2.1) and stray electric fields are properly taken into account. Since no fundamental limitations with respect to the achievable performance have been identified, an interesting question concerning the scalability of optical traps to larger numbers of ions arises. This is a particularly topical question given current efforts invested in scaling the capacities of platforms for quantum information processing based on ions, especially in light of the arising prospects of trapping two- and three-dimensional ion crystals without the detrimental effects of micromotion.

As will be discussed in the following, recent experiments have demonstrated that even comparatively simple experimental arrangements based on focused beam dipole traps can be used to trap linear chains of ions as depicted in Fig. 4.1a. A new major aspect that has to be considered as compared to traps for single ions is the mutual Coulomb repulsion of the constituents leading to additional electrostatic defocusing as illustrated in Fig. 4.1c. At the same time the presence of Coulomb interaction now provides access to the phonon modes of the trapped crystal, which is a prerequisite for applying trapped ions for quantum computation and simulation.

The corrections to the total trapping potential as given in Sect. 2.2 mainly consist of a generalized expression for the electrostatic contribution U_{dc} at the axial center of the trap, which now depends on position $\vec{r}_i = (x_i, y_i, z_i)$ and includes a defocusing potential stemming from the Coulomb repulsion between the ions U_{coul}:

$$U_{el}(\vec{r}_i) = U_{dc}(\vec{r}_i) + U_{coul}(\vec{r}_i). \tag{4.1}$$

Consequently, the approximated curvature of the electrostatic potential $U_{el}(\vec{r}_i)$ reads $\tilde{\omega}_{x,el}^2(z_i^0) = \tilde{\omega}_{x,dc}^2 + \tilde{\omega}_{x,coul}^2(z_i^0)$ for ion i in the weakest confined direction x. With

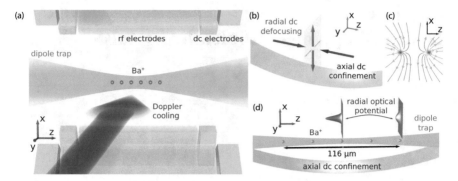

Fig. 4.1 Schematic of the experimental setup used for optical trapping of Coulomb crystals. (**a**) In the experiments, crystals containing up to six Barium ions, with lengths of up to 135 µm (5 ions: 116 µm) were prepared and optically trapped. (**b**) Electrostatic contributions to the effective potential due to axial dc confinement (gray surface). The defocusing mainly affected the x direction (red arrow) and was negligible in the y direction (thin gray arrow). (**c**) Additional defocusing in the radial directions due to ions' mutual Coulomb interaction. (**d**) The dipole traps were produced by focused Gauss beams (Rayleigh lengths $z_R^{VIS} = 40$ µm and $z_R^{NIR} = 74$ µm). Therefore, the radial optical potential (blue surfaces) for the outer ions was significantly lower and shallower than for the center ion. Taken from [1]

the discussed extensions of the potential model, the total radial potential energy for ion i at the axial position z_i^0 is then given by:

$$U_{tot}(x_i, y_i, z_i^0) = U_{opt}^{VIS}(x_i, y_i, z_i^0) + U_{el}(x_i, y_i, z_i^0), \qquad (4.2)$$

where the radial optical potential $U_{opt}^{VIS}(x_i, y_i, z_i^0)$ experienced by ion i also depends on its axial position z_i^0. This results in a trapping potential that is weakest for the outer ions as outlined in Fig. 4.2 and illustrates that the expected length scale for trappable Coulomb crystals will be set by the divergence of the employed Gaussian beam, and intuitively should be on the order of the Rayleigh range z_R.

A consequence of the position dependence of the effective potential is the increased sensitivity of the trapping performance to the alignment of the dipole trap beam with the z-axis of the Paul trap and to the beam quality. The latter can significantly deviate from the ideal radial profile of a Gaussian mode away from the focus. In addition, the contributions of stray electric fields have been neglected in the description of the effective potential. Furthermore, the techniques used for stray field compensation are most effective for linear electric fields measured at the center of the crystal, whereas higher order contributions or spatial field distributions require modified methods for detection, such as the concepts proposed in the following chapter, and a suitable configuration of compensation electrodes. These complications can lead to an effective reduction of trap depth for ions close to $\sim \pm z_R$.

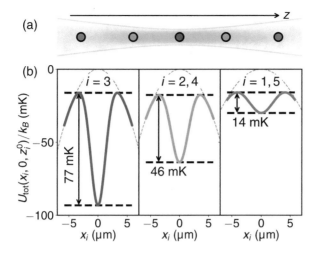

Fig. 4.2 (a) Illustration of a five ion crystal and (b) the effective potentials experienced by ions at positions z_i^0. The ions denoted with indices $i = 1 \ldots 5$ (blue, orange, and red circles) are exposed to different optical potentials (assuming a Gaussian dipole trapping beam with power $P_{opt}^{VIS} = 8$ W and waist radius $w_x^{VIS}(z = 0) = 2.6\,\mu$m) depending on their axial position along z. The electrostatic potential $U_{el}(\vec{r}_i)$ (dashed gray line) is approximated as a quadrupole potential and combines the contributions of an external electrostatic potential $U_{dc}(\vec{r}_i)$ (maximum defocusing along x-axis) and the repulsive Coulomb interaction $U_{coul}(\vec{r}_i)$ between the ions. The effective potential is calculated at the position of the ions with $i = 3$ (red), $i = 2, 4$ (orange) and $i = 1, 5$ (blue). Taken from [1]

Nonetheless, the results of recent experiments depicted in Fig. 4.3 show that Coulomb crystals with up to 6 ions can be trapped optically using essentially the protocol developed for single ion trapping outlined in Sect. 2.2. The used potential provided a maximal trap depth of ~ 100 mK $\times k_B$ generated by a focused Gaussian beam at a wavelength of 532 nm, with a power of 10 W, a waist radius of $2.6\,\mu$m, and a corresponding Rayleigh range of $40\,\mu$m. The length of the trapped chains reaches approximately $135\,\mu$m, which is on the order of $2z_R$ and is in agreement with the qualitative expectation based on the divergence of the Gaussian beam.

Interestingly, ions that are not addressed by laser cooling, such as barium isotopes other than ^{138}Ba$^+$ which appeared as dark gaps in the crystal, can be cooled sympathetically and are still trappable. In this case, they can act as markers witnessing a potential change of Coulomb ordering during the optical trapping attempts, as depicted in Fig. 4.3. Future applications, e.g., in the realm of quantum simulations with ions might build upon this property to realize optically trapped ion mixtures featuring a rich and in principle individually configurable phonon spectrum. In the investigated case, the probability for obtaining an unchanged configuration after a random redistribution of ions, i.e. due to melting and subsequent recooling, was found to be 9×10^{-10} for typical experimental realizations with four ion crystals. However, this result is insufficient for concluding that the trapped chains are actually

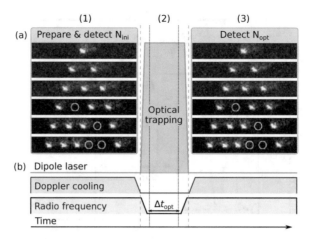

Fig. 4.3 Persistence of Coulomb order for an increasing number of optically trapped ions. (**a**) Fluorescence images of Coulomb crystals with $N_{ini} = 1 \dots 6$ Ba$^+$ ions are recorded before and after optical trapping of N_{opt}. For $N_{ini} \geq 4$, the gaps marked by orange circles reveal the presence of dark ions which appear at initial random lattice sites after Doppler and sympathetic cooling. (**b**) The experimental protocol (not to scale) consists of three steps: (1) we detect the initial configuration and ion number N_{ini} while Doppler cooling the ions; (2) the ions are transferred into the dipole trap by turning off the rf field and cooling lasers for the optical trapping duration Δt_{opt}, keeping the electrostatic potential; (3) we again detect the number and final configuration of all remaining ions in the rf trap. An intermittent gaseous phase followed by recrystallization or enhanced diffusion should be observable with high fidelity via changes of the positions of the dark ions within the crystal. Taken from [1]

crystals, since it is unclear whether the temperature of the trapped ion structure remained below the critical temperature $T_c \approx 50$ mK for the transition between a crystalline phase and a plasma.

4.1 Temperature of Ions in a Dipole Trap

More stressable indications for the survival of Coulomb crystals in the dipole trap can be obtained by measuring the temperature of the sample during the optical trapping phase. The corresponding technique discussed in Sect. 2.2 in principle is also applicable for more than a single ion and the results of such experiments are depicted in Fig. 4.4. However, additional contributions from the Coulomb repulsion have to be taken into account, and the effective potential has to be calculated for each ion individually. Since p_{opt} now denotes the survival probability of the entire ion ensemble, the extracted temperatures were obtained by calculating the effective potential at the position of the outermost ion experiencing the weakest confinement. The resulting temperatures were found to be consistent with the initial values of ≈ 1 mK at the end of the preparation phase, except for the case of four and five

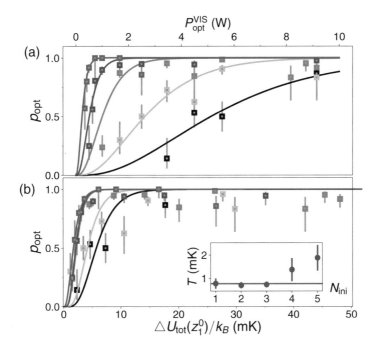

Fig. 4.4 (a) To obtain the temperature of ion chains, the trapping probability p_{opt} was measured for $N_{ini} = 1 \ldots 5$ ions as a function of optical power in the dipole trap P_{opt}^{VIS} (red, blue, green, yellow, and black squares), and the data was fitted the radial-cutoff model [2] (solid lined). Using the single-particle model resulted in a larger apparent temperature. (b) Analysis of the same data using an effective trap depth at locations of the outermost ions $\Delta U_{tot}(z_{1,N_{ini}}^0)$, accounting for Coulomb repulsion between ions and electrostatic defocusing. Fits assuming the modified radial-cutoff model shown as solid lines, yielding temperatures of $T_{N_{ini}} = (0.7 \pm 0.1)$ mK ($N_{ini} \leq 3$), $T_4 = (1.3 \pm 0.5)$ mK and $T_5 = (1.8 \pm 0.5)$ mK (inset). Taken from [1]

ion chains. The apparent increase of temperature for larger ion chains could be due to the complications arising from increasing sensitivity to alignment, beam shape and stray electric fields, which are not accounted for in the model. Since all measured values were found to be at least one order of magnitude below T_c, these results strongly suggest that the prepared ion chains remained crystallized during the optical trapping process.

4.2 Accessing Vibrational Modes of Coulomb Crystals in Optical Traps

A direct signature of the crystalline phase of the trapped ions can be observed by performing an analysis of the vibrational spectrum. This can be achieved for example by periodically modulating the axial electrostatic potentials at a frequency

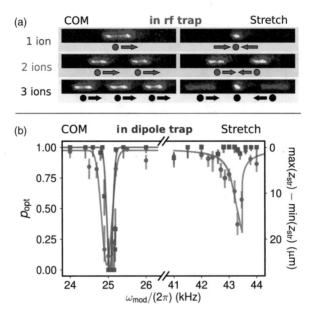

Fig. 4.5 (a) Fluorescence images of ions in the Paul trap showing resonances excited by applying a modulation of the electrostatic axial confinement (distance between two ions is $43\,\mu m$) at modulation frequencies $\omega_{mod} = \omega_{ax}^{COM}$ (axial motion, $N_{ini} = 1$), and at $\omega_{ax}^{str} = \sqrt{3}\,\omega_{ax}^{COM}$ ($N_{ini} = 2, 3$, out-of-phase motion). (b) Optical trapping probability for one (blue squares) and two (red circles) ion(s) in the dipole trap, as a function of the modulation frequency $\omega_{mod}/2\pi$ of the axial electrostatic potential. A drop in p_{opt} at ω_{ax}^{COM} is observed for both $N_{ini} = 1$ and $N_{ini} = 2$ (solid lines: fits to data). The resonance at $\omega_{ax}^{str} = 2\pi \times (43.3 \pm 0.15)\,kHz \approx \sqrt{3}\,\omega_{ax}^{COM}$ was observed only for $N_{ini} = 2$. In the case of the stretch mode, we modulate the harmonic confinement to excite differential motion between the ions. The electric field amplitude at the equilibrium positions of the ions amounts to $|E| = (1.8 \pm 0.1)\,mV/m$ for which we numerically simulate the out-of-phase motion (no free parameters). The amplitude of the stretch mode is given by $max(z_{str}(t)) - min(z_{str}(t))$ (solid red line, axis on the right-hand side). Taken from [1]

ω_{mod}, effectively exciting the trapped ions on the corresponding normal modes [3] whenever the modulation occurs at their characteristic frequencies. For laser cooled ions confined in a Paul trap this resonant excitation manifests itself in a blurring of the ions as shown in Fig. 4.5a for the case of the center of mass (COM) and stretch modes of up to three ions found at $\omega_{ax}^{COM} \approx 2\pi \times 25\,kHz$ and $\omega_{ax}^{str} \approx 2\pi \times 43\,kHz$ (two ion case), respectively. In order to prove the existence of these modes for optically trapped ion strings, this method has to be applied during the optical trapping phase. In a recent realization within a dipole trap operated at 1064 nm, samples of one and two ions trapped for $\tau = 10\,ms$ were exposed to such periodic excitations. These experiments revealed sharp resonances signaled by a drop of trapping probability p_{opt} at $\omega_{mod} \approx \omega_{ax}^{COM}$ for one ($N_{ini} = 1$) and two ($N_{ini} = 2$) ions, whereas an additional resonance around ω_{ax}^{str} only could be observed in experiments carried out with two ion chains, as illustrated in Fig. 4.5b.

These findings confirmed that the crystals prepared in the Paul trap survived the transfer into the dipole trap, allowing for access and manipulation of their normal modes [1] during the optical trapping phase.

4.3 Limitations and Prospects

While the first investigations highlighted in this chapter show that optical traps can be used as a new platform to investigate ion Coulomb crystals in absence of any radiofrequency fields, an important question in view of future applications pertains to the expected limitations with respect to trappable ion numbers and crystal configurations. To this end, all experiments carried out so far both with single ions as well as with Coulomb crystals provide no evidence for a fundamental limitation of the applied method. This is to say that based on the experimental evidence available at the moment, ions can be trapped and manipulated using the same methods employed in neutral atom experiments when accounting for the contributions from mutual Coulomb interaction and electrostatic potentials, with comparable performance regarding trapping and coherence times. As shown in the case of linear ion crystals, the maximal number of trappable ions is consistent with physical constraints imposed by the chosen geometry of the dipole trap. Whereas focused beam dipole traps provide several advantages for trapping single ions, such as a comparably simple and robust experimental setup, specific applications may benefit from more adapted arrangements employing the superposition of several optical fields, which are well-understood and routinely used in neutral atom experiments [4, 5]. Nonetheless, the described apparatus could be extended and modified to trap higher dimensional crystals with up to 10 ions, using standard techniques and commercially available laser sources.

Specifically, several intriguing ideas formulated in recent theoretical works in the field of structural quantum phase transitions [6–9] and quantum simulations [10, 11] may be within reach already with currently available ion numbers. Another recently emerged field of research where the described techniques may help accessing a new experimental regime is the study of ion–atom interactions at low and ultralow temperatures [12, 13]. So far, even the most promising approaches for reaching the quantum regime of interaction based on exploiting a favorable combination of single Yb^+ ions confined in a conventional rf trap and ultracold Li atoms [14, 15] may be effective for a single ion immersed in the atom cloud, while their extension to linear ion chains is likely to be most challenging.

References

1. J. Schmidt, A. Lambrecht, P. Weckesser, M. Debatin, L. Karpa, T. Schaetz, Optical trapping of ion Coulomb crystals. Phys. Rev. X **8**(2), 021028 (2018). ISSN 2160-3308. https://doi.org/10.1103/PhysRevX.8.021028. http://arxiv.org/abs/1712.08385

2. C. Schneider, M. Enderlein, T. Huber, S. Dürr, T. Schaetz, Influence of static electric fields on an optical ion trap. Phys. Rev. A **85**, 013422 (2012). https://doi.org/10.1103/PhysRevA.85. 013422
3. D.J. Wineland, Nobel lecture: superposition, entanglement, and raising Schrödinger's cat. Rev. Mod. Phys. **85**(3), 1103–1114 (2013). https://doi.org/10.1103/revmodphys.85.1103
4. I. Bloch, J. Dalibard, W. Zwerger, Many-body physics with ultracold gases. Rev. Mod. Phys. **80**, 885–964 (2008). https://doi.org/10.1103/RevModPhys.80.885
5. O. Morsch, M. Oberthaler, Dynamics of Bose-Einstein condensates in optical lattices. Rev. Mod. Phys. **78**(1), 179–215 (2006). ISSN 00346861. https://doi.org/10.1103/RevModPhys.78. 179
6. J.D. Baltrusch, C. Cormick, G. De Chiara, T. Calarco, G. Morigi, Quantum superpositions of crystalline structures. Phys. Rev. A **84**, 063821 (2011). https://doi.org/10.1103/PhysRevA.84. 063821
7. J.D. Baltrusch, C. Cormick, G. Morigi, Quantum quenches of ion Coulomb crystals across structural instabilities. Phys. Rev. A **86**, 032104 (2012). https://doi.org/10.1103/PhysRevA.86. 032104
8. E. Shimshoni, G. Morigi, S. Fishman, Quantum Zigzag transition in ion chains. Phys. Rev. Lett. **106**, 010401 (2011). https://doi.org/10.1103/PhysRevLett.106.010401.
9. E. Shimshoni, G. Morigi, S. Fishman, Quantum structural phase transition in chains of interacting atoms. Phys. Rev. A **83**, 032308 (2011). https://doi.org/10.1103/PhysRevA.83. 032308
10. R. Nath, M. Dalmonte, A.W. Glaetzle, P. Zoller, F. Schmidt-Kaler, R. Gerritsma, Hexagonal plaquette spin-spin interactions and quantum magnetism in a two-dimensional ion crystal. New J. Phys. **17**(6), 065018 (2015). ISSN 1367-2630. https://doi.org/10.1088/1367-2630/17/ 6/065018
11. U. Bissbort, D. Cocks, A. Negretti, Z. Idziaszek, T. Calarco, F. Schmidt-Kaler, W. Hofstetter, R. Gerritsma, Emulating solid-state physics with a hybrid system of ultracold ions and atoms. Phys. Rev. Lett. **111**, 080501 (2013). https://doi.org/10.1103/PhysRevLett.111.080501
12. A. Härter, J. Hecker Denschlag, Cold atom–ion experiments in hybrid traps. Contemp. Phys. **55**(1), 33–45 (2014). https://doi.org/10.1080/00107514.2013.854618
13. M. Tomza, K. Jachymski, R. Gerritsma, A. Negretti, T. Calarco, Z. Idziaszek, P.S. Julienne, Cold hybrid ion-atom systems. Rev. Mod. Phys. **91**, 035001 (2019)
14. M. Cetina, A.T. Grier, V. Vuletić, Micromotion-induced limit to atom-ion sympathetic cooling in Paul traps. Phys. Rev. Lett. **109**, 253201 (2012). https://doi.org/10.1103/PhysRevLett.109. 253201
15. H.A. Fürst, N.V. Ewald, T. Secker, J. Joger, T. Feldker, R. Gerritsma, Prospects of reaching the quantum regime in Li-Yb + mixtures. J. Phys. B Adv. At. Mol. Opt. Phys. **51**(19), 195001 (2018). ISSN 0953-4075. https://doi.org/10.1088/1361-6455/aadd7d

Chapter 5
Summary and Outlook

Over the past years optical traps for ions have undergone a development from first conceptual realizations of single ion traps in proof-of-principle experiments to a point where they can be used as a novel platform featuring a number of unique properties including complete isolation from radiofrequency fields, nanoscale potentials, and state-selectivity. On the one hand, all experiments carried out to date clearly show that the coupling of ions to electric fields makes it necessary to account for additional contributions to the trapping potentials as compared to neutral atoms, many of them leading to a reduced trap depth. Nonetheless, the techniques that have been developed and applied with great success in the case of neutral atoms, most notably the use of far-off-resonance traps, can also be applied for ions in dipole traps and gives rise to a comparable reduction of detrimental off-resonant scattering and the related enhancement of the trapping performance. This is exemplarily evident in the observed increase of ion lifetime from initially milliseconds to timescales on the order of several seconds achievable today. On the other hand, facing the additional challenges in comparison to neutral atom traps has also led to the development and improvement of numerous methods for manipulating ions. For example, the restrictions arising from the presence of stray electric fields have been tackled by improving upon known methods for their detection and compensation to levels below 10 mV/m.

Of course, adapting optical traps to ions will not remain without additional challenges and drawbacks that have to be addressed in the future. For instance, the Coulomb repulsion between ions in a crystal poses unavoidable restrictions with respect to the required optical powers. This puts a limit on the minimal distances and the related coupling rates. Similarly, the stray field compensation techniques successfully applied so far have not yet been adopted to the case of large ion strings or two- and three-dimensional Coulomb crystals. Nevertheless, an important overarching result of previous work is that so far no fundamental limitations have been encountered, and that the main restrictions to the performance arise from technical or avoidable aspects, such as trap geometry or unfavorable

L. Karpa, *Trapping Single Ions and Coulomb Crystals with Light Fields*, SpringerBriefs in Physics, https://doi.org/10.1007/978-3-030-27716-1_5

branching ratios for specific ion species. As such, optical traps for ions hold promise of continued improvement, in particular with respect to achievable ion numbers, trapping, and coherence time.

Exploiting the features granted by the combination of optical traps and ions at the currently available level makes it feasible to approach a novel class of experiments with ions, Coulomb crystals, and ion–atom mixtures in the fields of quantum simulations [1], quantum information processing [2], ultracold ion–atom interactions [3], and metrology. In the following, I will highlight some novel applications that could be realized in the near future drawing on the new capabilities provided by optical trapping.

5.1   State-Selective Potentials

An intriguing application of optical traps for ion crystals is the capability to directly realize state-selective potentials. In experiments with neutral atoms this is a well-established and common feature. Radiofrequency-based ion traps or Penning traps on the other hand are very difficult or potentially impossible to configure in a way that provides different confinements depending on the electronic state of individual ions. This is due to the fact that the confinement of ions in rf traps stems from the interaction of the charged nuclei to radiofrequency fields, whereas the interaction of the ions to the electromagnetic fields arises from their coupling to the electrons. While rf fields enable very deep trapping potentials for ions on the order of $10^4 K * k_B$, in general these traps are independent of the ions' electronic states, although new promising approaches for the realization of state-selectivity, e.g. based on the properties of Rydberg ions [4–6], have been brought forward recently. At the same time the capability to exert state-selective forces is crucial for implementing quantum computation algorithms, quantum walks, or quantum simulations with ions, and several methods utilizing optical forces within a common state-insensitive potential of an rf trap have been developed to this end [7–12]. As discussed in the previous chapters, optical dipole traps are intrinsically ideally suitable for this task, since in analogy to the case of neutral atoms, optical forces come about from light shifts of the electronic states of the ions. Provided a suitable configuration of the electronic level structure exists, this readily allows to individually and controllably manipulate the trapping potentials of individual ions within a crystal. For example, Ba^+ ions feature metastable D levels located between the ground and excited states. Ions prepared in such metastable states experience a different potential depending on the wavelength of the dipole trap. In particular, an optical trap operated at a wavelength of 532 nm induces an attractive potential in the electronic ground state, but ions in one of the D levels are repelled by the same optical field. Individual shelving of an arbitrary sub-set of ions in a Coulomb crystal can then be used to remove such ions with very high fidelity.

Another novel application of these properties opens up when a far-off-resonant red-detuned dipole trap is used. This is the case for a trap operated at a wavelength of

1064 nm, where both the S and the D states are trappable, but the trap depths differ by about a factor of 4. This can be used to realize entanglement between internal and external degrees of freedom as well as in experiments investigating structural quantum phase transitions, e.g. by preparing ions in a superposition of S and D states [13–16].

Similarly, the state-selectivity of optical traps is a key feature that could enable a theoretically proposed possibility of realizing fracton models by exploiting analogies between the relevant properties of fractons and elasticity theory of two-dimensional quantum crystals [17]. For example, the fractons, dipoles, and gauge modes encountered in tensor gauge theory can be mapped onto disclinations, dislocations, and phonons of two-dimensional crystals described by elasticity theory. Ion Coulomb crystals are potentially a highly suitable platform for observing this mapping, provided that the topological defects can be created, for example by selective removal of a row (M. Pretko, L. Radzihovsky, private communication (2018)). This can be achieved by shelving Yb^+ or Ba^+ ions into metastable D manifolds while keeping the whole crystal in a dipole trap with a wavelength of 532 nm.

So far, all realizations of optical trapping without the assistance of radiofrequency fields have been carried out in macroscopically spaced three-dimensional traps. This is mainly a consequence of the large beam diameters used to obtain small beam waists of the dipole traps on the order of a few μm which are highly beneficial due to the correspondingly large curvatures of the optical potential. However, with the rapid developments achieved in the field of planar ion traps, similar results could be obtained in the future by utilizing traps with integrated optical components capable of focusing to the μm scale as demonstrated in recent work at the Massachusetts Institute of Technology [18]. Such a realization would allow to interface ideas behind QCCD-type trapping architectures designed for applications in QIP with dipole traps [19], providing rf-free interaction regions that can be shifted away from the surface, thereby mitigating or avoiding anomalous heating. The latter is considered as one of the main obstacles for planar trap miniaturization and scalability of ion-based QIP platforms. Along similar lines, optical traps could be used to measure heating rates in absence of rf fields, giving important insight into this active area of research complementary to currently employed techniques. The fundamental compatibility of optical potentials and planar ion trap geometries has already been demonstrated in experiments combining such traps with one-dimensional standing wave potentials [20–22].

5.2 Higher Dimensional Coulomb Crystals in Optical Lattices

The only demonstration of optical trapping of more than one ion so far was achieved for a linear ion string within a hybrid trap where the radial confinement was still provided by rf fields [20–22], which are detrimental in applications involving ion–atom collisions [3]. More recently all-optical trapping of ion crystals in the absence of rf fields has been observed in focused beam far-off-resonance traps (FORT) [23]. Scaling of the system to 2d and 3d requires a means to overcome the limitations in ion numbers arising from the finite range of the axial trapping potential, as discussed in Sect. 4.3. This could be achieved in the near future by overlapping two or more optical fields to form a standing wave, as outlined in the following.

The most simple implementation would be based on an adoption of techniques demonstrated in [24] by utilizing the optical potential of a one-dimensional standing wave. The latter would be axially aligned with a ring-type Paul trap configured to provide *radial electrostatic* and *axial rf* confinement. The advantage of this configuration consists in the avoidance of loss along the radial directions. A planar Coulomb crystal can be shaped and oriented in the radial plane by choosing parameters such that the curvature of the radial electrostatic potential is substantially weaker than the axial curvature of the rf potential. The Coulomb repulsion between the trapped ions is then absorbed at the expense of strong axial defocusing. The latter can now be overcome by the much steeper gradient of the intensity achievable in a standing wave or by superimposing two optical fields with wavelengths λ and $\lambda/2$ to generate a superlattice with an axial intensity envelope compensating the electrostatic defocusing.

In the concrete setting described in the previous chapters, one might make use of laser sources operating at the wavelengths of 1064 nm and, after frequency doubling, 532 nm, as employed in a recent work with neutral ^6Li atoms [25]. These two optical fields then can be arranged to produce aligned standing waves such that at the axial center position the depth of the combined optical potential in the adjacent potential wells is initially increasing for larger distances away from the center cite. This could provide robustness against stray electric fields, which are increasingly difficult to compensate for larger crystals. For example, stray field displacement by 5 μm only slightly changes the potential in a lattice site, but moves the ion out of the capture range of a focused beam dipole trap, as illustrated in Fig. 5.1. Building upon the concepts introduced in recent theoretical proposals [13–16], such a configuration would also be suitable for studies of structural quantum phase transitions between a 2d ion crystal and a 3d crystal by means of tunneling of the center ion, or for creating entangled states between internal degrees of freedom and structural crystal phases.

More complex and versatile geometries can be realized by adopting the concepts successfully applied for neutral atoms making use of several laser beams and controlling their respective degrees of freedom. For instance, superpositions of three suitably aligned beams give complex structures such as Kagomé lattices. For neutral atoms these ideas have been extended even further, now allowing the construction of arbitrarily configurable three-dimensional micro-trap structures with more than a hundred individual sites, as demonstrated very recently [26]. In principle, all these techniques can also be applied to the case of higher dimensional Coulomb crystals. In the future, this novel experimental approach should provide a platform for

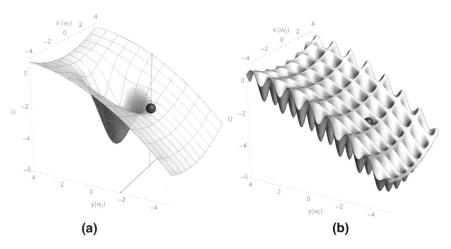

(a)   **(b)**

Fig. 5.1 (**a**) Illustration of the effective potential U of a dipole laser field focused to a waist of w_0 in presence of both curvature due to the electrostatic field of the trap electrodes (additional harmonic confinement along the x-axis, repulsion along y) and a finite stray electric field (tilt of the potential along y), leading to reduced trap depths. An ion displaced beyond the edge of the reduced optical capture region at $y \approx -2\,w_0$ gets lost due to the repulsive component of the electrostatic potential. (**b**) Corresponding potential of a 2D optical lattice same electrostatic potential as in (**a**), obtained by superimposing two orthogonal light fields with beam waists of 5 w_0 (optical power same as in (**a**)). Here, the optical forces are increased by $\sim w_0/(\lambda/2)$ (ratio of gradients of intensity), such that displacement due to the same stray electric field is suppressed and the ion remains trapped

investigating a to date unexplored interplay of Coulomb order and optical periodicity beyond the one-dimensional case that already has enabled to disseminate friction models in Frenkel–Kontorova-type systems at an unprecedented level of accuracy and detail [21, 22].

Novel and intriguing applications of the interaction between ion crystals and tailored optical potentials may also arise in the context of quantum simulations. For example, two-dimensional Coulomb crystals with a comparably small number of constituents have been recently proposed as experimental platforms for simulating spin-spin interactions on a hexagonal plaquette and quantum magnetism [1], where optical forces are used to tailor the phonon-mode spectrum. In the future, trapping of ions in designated sites of optical lattices may enable even more refined control allowing for individual shaping of the local phonon modes.

5.3 Metrological Applications

The methods for overlapping ions with optical fields described in the previous chapters, in particular in Sect. 2.2, may also be useful in the realm of metrological applications. For example, ion crystals aligned with standing waves could be

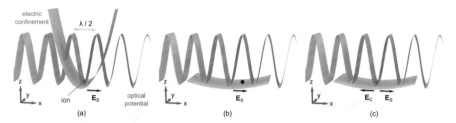

Fig. 5.2 Schematic of the procedure for compensating a stray electric field $\mathbf{E}_S$ based on the scheme in reference [27]. (**a**) Laser cooled ions placed in the nodes of the standing wave experience virtually no ac Stark shift and scatter light at a rate R_0. Displacements from the null of the (rf) electrostatic (pseudo)potential are small due to conventional pre-compensation and the steepness of the trap. (**b**) As the electric trapping potential is lowered the ion gets displaced to positions where the ac Stark shift Δ_{ac} is considerable, shifting it out of resonance and reducing the scattering rate to $\sim R_0/\Delta_{ac}^2$. Here the sensitivity to displacements due to $\mathbf{E}_S$ is enhanced to a fraction of $\lambda/2$ compared to the currently best compensation method with a characteristic length scale determined by the waist w_0 of the focused dipole trap beam [27] ($w_0 = 4\,\mu$m). (**c**) Compensation is detected by measuring fluorescence while applying a compensation field $\mathbf{E}_C$ such that the ion is moved back to the node of the standing wave

used for sensing of external fields. To this end, in the simplest case individual positions could be mapped to spatially resolved fluorescence patterns, as illustrated for one dimension in Fig. 5.2. This straight-forward method is expected to allow for detecting external field gradients and their evolution simultaneously with high spatial and temporal resolution, as determined by the spacing between ions of typically a few $10\,\mu$m and the available detection sensitivity for stray electric fields of $\delta E_S \approx 10\,$mV/m after integration times of about $10\,$s. In more advanced scenario the obtained sensitivity could then be enhanced even further by employing spectroscopic methods for measuring the locally induced light shifts in the energy levels of individual ions rather than by monitoring their fluorescence. Three-dimensional maps can then be obtained by observing 3d Coulomb crystals either exposed to a 3d optical lattice or a set of orthogonal 1d standing waves switched on in sequence.

Another exciting approach is the combination of optical trapping techniques with surface electrode traps, either by interfacing Coulomb crystals with optical cavities [20] or by adapting recently developed methods for direct integration of optical components [18] with planar traps. These concepts could be used and extended in the near future to realize novel experimental arrangements for measuring heating rates close to electrode surfaces in absence of radiofrequency fields or, more conservatively, for creating optical patterns for sensing of electric field distributions.

References

1. R. Nath, M. Dalmonte, A.W. Glaetzle, P. Zoller, F. Schmidt-Kaler, R. Gerritsma, Hexagonal plaquette spin–spin interactions and quantum magnetism in a two-dimensional ion crystal. New J. Phys. **17**(6), 065018 (2015). ISSN 1367-2630. http://dx.doi.org/10.1088/1367-2630/17/6/065018

2. J.I. Cirac, P. Zoller, A scalable quantum computer with ions in an array of microtraps. Nature **404**(6778), 579–581 (2000). http://dx.doi.org/10.1038/35007021

3. M. Cetina, A.T. Grier, V. Vuletić, Micromotion-induced limit to atom-ion sympathetic cooling in Paul traps. Phys. Rev. Lett. **109**, 253201 (2012). https://doi.org/10.1103/PhysRevLett.109.253201

4. T. Feldker, P. Bachor, M. Stappel, D. Kolbe, R. Gerritsma, J. Walz, F. Schmidt-Kaler, Rydberg excitation of a single trapped ion. Phys. Rev. Lett. **115**, 173001 (2015). https://doi.org/10.1103/PhysRevLett.115.173001

5. G. Higgins, W. Li, F. Pokorny, C. Zhang, F. Kress, C. Maier, J. Haag, Q. Bodart, I. Lesanovsky, M. Hennrich, Single strontium Rydberg ion confined in a Paul trap. Phys. Rev. X **7**, 021038 (2017). https://doi.org/10.1103/PhysRevX.7.021038

6. G. Higgins, F. Pokorny, C. Zhang, Q. Bodart, M. Hennrich, Coherent control of a single trapped Rydberg ion. Phys. Rev. Lett. **119**, 220501 (2017). https://doi.org/10.1103/PhysRevLett.119.220501

7. D.J. Wineland, Nobel lecture: superposition, entanglement, and raising Schrödinger's cat. Rev. Mod. Phys. **85**(3), 1103–1114 (2013). https://doi.org/10.1103/revmodphys.85.1103

8. D. Leibfried, R. Blatt, C. Monroe, D. Wineland, Quantum dynamics of single trapped ions. Rev. Mod. Phys. **75**, 281–324 (2003). https://doi.org/10.1103/RevModPhys.75.281

9. C. Monroe, D.M. Meekhof, B.E. King, S.R. Jefferts, W.M. Itano, D.J. Wineland, P. Gould, Resolved-sideband Raman cooling of a bound atom to the 3D zero-point energy. Phys. Rev. Lett. **75**(22), 4011–4014 (1995). ISSN 1079-7114. https://doi.org/10.1103/PhysRevLett.75.4011. http://www.ncbi.nlm.nih.gov/pubmed/10059792

10. R. Blatt, C.F. Roos, Quantum simulations with trapped ions. Nat. Phys. **8**(4), 277–284 (2012). https://doi.org/10.1038/nphys2252

11. A. Friedenauer, H. Schmitz, J.T. Glückert, D. Porras, T. Schaetz, Simulating a quantum magnet with trapped ions. Nat. Phys. **4**(10), 757–761 (2008). ISSN 1745-2473. http://dx.doi.org/10.1038/nphys1032

12. R. Blatt, D. Wineland, Entangled states of trapped atomic ions. Nature **453**(7198), 1008–1015 (2008). ISSN 0028-0836. https://doi.org/10.1038/nature07125. http://www.ncbi.nlm.nih.gov/pubmed/18563151

13. J.D. Baltrusch, C. Cormick, G. De Chiara, T. Calarco, G. Morigi, Quantum superpositions of crystalline structures. Phys. Rev. A **84**, 063821 (2011). https://doi.org/10.1103/PhysRevA.84.063821

14. J.D. Baltrusch, C. Cormick, G. Morigi, Quantum quenches of ion Coulomb crystals across structural instabilities. Phys. Rev. A **86**, 032104 (2012). https://doi.org/10.1103/PhysRevA.86.032104

15. E. Shimshoni, G. Morigi, S. Fishman, Quantum Zigzag transition in ion chains. Phys. Rev. Lett. **106**, 10401 (2011). https://doi.org/10.1103/PhysRevLett.106.010401

16. E. Shimshoni, G. Morigi, S. Fishman, Quantum structural phase transition in chains of interacting atoms. Phys. Rev. A **83**, 032308 (2011). https://doi.org/10.1103/PhysRevA.83.032308

17. M. Pretko, L. Radzihovsky, Fracton-elasticity duality. Phys. Rev. Lett. **120**(19), 195301 (2018). ISSN 0031-9007. https://doi.org/10.1103/PhysRevLett.120.195301. https://link.aps.org/doi/10.1103/PhysRevLett.120.195301

18. K.K. Mehta, C.D. Bruzewicz, R. McConnell, R.J. Ram, J.M. Sage, J. Chiaverini, Integrated optical addressing of an ion qubit. Nat. Nanotechnol. **11**(12), 1066–1070 (2016). ISSN 1748-3387. https://doi.org/10.1038/nnano.2016.139. http://www.nature.com/articles/nnano.2016.139

19. D. Kielpinski, C. Monroe, D.J. Wineland, Architecture for a large-scale ion-trap quantum computer. Nature **417**(6890), 709–711 (2002). https://doi.org/10.1038/nature00784. http://dx.doi.org/10.1038/nature00784

20. L. Karpa, A. Bylinskii, D. Gangloff, M. Cetina, V. Vuletić, Suppression of ion transport due to long-lived subwavelength localization by an optical lattice. Phys. Rev. Lett. **111**, 163002 (2013). https://doi.org/10.1103/PhysRevLett.111.163002

21. A. Bylinskii, D. Gangloff, V. Vuletic, Tuning friction atom-by-atom in an ion-crystal simulator. Science **348**(6239), 1115–1118 (2015). https://doi.org/10.1126/science.1261422

22. D. Gangloff, A. Bylinskii, I. Counts, W. Jhe, V. Vuletic, Velocity tuning of friction with two trapped atoms. Nat. Phys. **11**(11), 915–919 (2015). ISSN 1745-2473. http://dx.doi.org/10.1038/nphys3459

23. J. Schmidt, A. Lambrecht, P. Weckesser, M. Debatin, L. Karpa, T. Schaetz, Optical trapping of ion coulomb crystals. Phys. Rev. X **8**(2), 021028 (2018). ISSN 2160-3308. https://doi.org/10.1103/PhysRevX.8.021028. http://arxiv.org/abs/1712.08385

24. M. Enderlein, T. Huber, C. Schneider, T. Schaetz, Single ions trapped in a one-dimensional optical lattice. Phys. Rev. Lett. **109**, 233004 (2012). https://doi.org/10.1103/PhysRevLett.109.233004

25. M. Tomza, K. Jachymski, R. Gerritsma, A. Negretti, T. Calarco, Z. Idziaszek, P.S. Julienne, Cold hybrid ion-atom systems, Rev. Mod. Phys. **91**, 035001 (2019)

26. D. Barredo, V. Lienhard, S. de Léséleuc, T. Lahaye, A. Browaeys, Synthetic three-dimensional atomic structures assembled atom by atom. Nature **561**(7721), 79–82 (2018). ISSN 0028-0836. https://doi.org/10.1038/s41586-018-0450-2. http://www.nature.com/articles/s41586-018-0450-2

27. T. Huber, A. Lambrecht, J. Schmidt, L. Karpa, T. Schaetz, A far-off-resonance optical trap for a Ba$^+$ ion. Nat. Commun. **5**, 5587 (2014)

Index

© The Author(s), under exclusive license to Springer Nature Switzerland AG 2019   47
L. Karpa, *Trapping Single Ions and Coulomb Crystals with Light Fields*,
SpringerBriefs in Physics, https://doi.org/10.1007/978-3-030-27716-1

Printed in the United States
By Bookmasters